THE AUTHENTICITY HOAX

THE AUTHENTICITY HOAX

How We Get Lost Finding Ourselves

ANDREW POTTER

HARPER

An Imprint of HarperCollins*Publishers*
www.harpercollins.com

For Stephanie, Marnie, and Matthew

Published in Canada in 2010 by McClelland & Stewart Ltd.

FIRST U.S. EDITION

Library of Congress Cataloging-in-Publication Data

Potter, Andrew.
  The authenticity hoax : how we got lost finding ourselves / Andrew Potter. — 1st ed.
      p.   cm.
  Includes bibliographical references and index.
  ISBN 978-0-06-125133-7
 1. Social evolution.   2. Mass society.   3. Authenticity (Philosophy)
4. Identity (Psychology)   I. Title.
  HM626.P68   2010
  306.09'0511—dc22

                                                      2009045459

10   11   12   13   14   OFF/RRD   10   9   8   7   6   5   4   3   2   1

But now being lifted into high society,
And having pick'd up several odds and ends
Of free thoughts in his travels for variety,
He deem'd, being in a lone isle, among friends,
That, without any danger of a riot, he
Might for long lying make himself amends;
And, singing as he sung in his warm youth,
Agree to a short armistice with truth.

    – Lord Byron

CONTENTS

THE JARGON OF AUTHENTICITY

IN THE SUMMER OF 2008, A TWENTY-EIGHT-YEAR-OLD FRENCH engineer from Brittany named Florent Lemaçon, his wife, Chloé, and their three-year-old son, Colin, embarked on what looked to be the trip of a lifetime. After quitting their jobs, the Lemaçons set sail from France in a boat into which they had poured their life savings, a restored yacht named the *Tanit*. Their ultimate destination was Zanzibar, an archipelago off the coast of Tanzania, and to help them sail around the clock, the Lemaçons had picked up another couple. As the *Tanit* and its crew left Egypt and headed down into the Indian Ocean, they spoke to a French frigate that strongly advised them to turn back from a journey that would take them into some of the most lawless, pirate-infested waters in the world.

The undaunted adventurers continued on their way, and over the weekend of April 4, 2009, they were seized by Somali pirates who were intent on taking their five hostages back to the mainland, where they would be harder to find and, hence, easier to ransom. After negotiations with the pirates broke down, French commandos launched a rescue operation during which four of the *Tanit* crew were rescued. Mr. Lemaçon himself was killed during the ensuing gunfight, perhaps by friendly fire as he tried to duck down into the yacht's cabin.

Why did they continue their voyage, despite being repeatedly warned about the dangers? On a blog the couple kept of their trip, the Lemaçons wrote: "The danger is there and has indeed become greater over the past months, but the ocean is vast. . . . The pirates must not be allowed to destroy our dream." And their dream, as they told everyone who would listen, was to protect their son, Colin, from the depraved elements of the modern world, especially the sterile government and its officious bureaucracy, the shallowness of the mass media, and the meaninglessness of consumer society and its destructive environmental impact. "We don't want our child to receive the sort of education that the government is concocting for us," Florent told a French newspaper. "We have got rid of the television and everything that seemed superfluous to concentrate on what is essential."

The story of a disillusioned young man looking for meaning outside the iron cage of modern life was a cliché even by the time Henry David Thoreau went off to Walden Pond, and Florent Lemaçon is not the first person to get himself killed while searching for a leaner and less complicated mode of existence. But there is something especially pathetic and pointless about this case, even discounting Florent and Chloé's outrageous decision to bring their young son on such a trip. Civilization has its drawbacks, but if there is one unambiguous good that it provides it is safety, security, and the rule of law. It is one thing to look to escape into nature, something else entirely to head deliberately into a lawless realm of high-seas piracy at a time when the papers were full of stories about ships being taken and crews being held hostage for millions of dollars in ransom. Only someone in the grip of a seriously misguided ideological quest could imagine that taking his family through the Gulf of Aden is a more "essential" form of existence, or a reasonable and virtuous alternative to the life of a well-paid professional in contemporary France.

Yet for all their recklessness, there is nothing remotely eccentric about what Florent and Chloé Lemaçon were searching for. The object of their desire, the "essential" core of life, is something called *authenticity*, and finding the authentic has become the foremost spiritual quest of our time. It is a quest fraught with difficulty, as it takes place at the intersection of some of our culture's most controversial issues, including environmentalism and the market economy, personal identity and consumer culture, and artistic expression and the meaning of life.

One widely accepted view is that it is impossible to build an authentic personal identity out of the cheap building blocks of consumer goods, while an essential part of living authentically involves treading softly upon the earth and leaving as small a footprint as possible. When it comes to personal fulfillment, many of us subscribe to the idea that the self is an act of artistic creation, and living a meaningful, creative life is impossible within the confines of the modern world. And so many of us seek a more authentic form of life outside of modernity or in opposition to it, a quest that is in many ways as old as the Romantic turn that arose in reaction to the Enlightenment.

Yet too often for comfort, the search for the authentic is itself twisted into just another selling point or marketing strategy, and once we appreciate the full implications of this, there is a real danger that cynicism will quickly set in: everyone is working an angle, everyone is looking to make a buck. Once we start down this path, it isn't long before we reach the same conclusion as Florent and Chloé Lemaçon – society is corrupt, commerce is alienating, and the whole system should be abandoned, if not completely destroyed.

We can tie ourselves into knots over this, but the fact is, the relationship between the stuff we buy and who we are, and the broader relationship among consumer culture, artistic vision, and the authentic self, is fraught with bad arguments and bad faith, and the usual themes and oppositions (between genuine needs and false wants, or between the shallowness of a branded identity and the depths of the true self) are too crude to be helpful.

We need a new approach, one that takes seriously our desire for an authentic, meaningful, ecologically sensible life, but that recognizes that the market economy, along with many other aspects of the modern world, are not evils, even necessary ones, but are instead a rich and vibrant source of value that we would not want to abandon, even if it were possible.

We live in a world increasingly dominated by the fake, the prepackaged, and the artificial. Whichever way we turn we are beset by outrageous advertising, lying politicians, and fraudulent memoirists. Some of us live in cookie-cutter suburban developments, others in gentrified urban neighborhoods almost indistinguishable from theme parks. We eat barely nutritious fast food, watch scripted "reality" television shows, and take prepackaged vacations complete with prepackaged memories. Meanwhile, we continuously find refuge on the Internet, where we spend enormous amounts of time hanging out on Facebook messaging our "friends" or wandering around virtual environments like Second Life or World of Warcraft, interacting with the avatars of people we've never actually met and couldn't recognize if we did.

But if we look around, there are intimations of a growing backlash as the demand for the honest, the natural, the real – that is, the authentic – has become one of the most powerful movements in contemporary life. In reaction against the isolation, shallowness, and alienation of everyday life, people everywhere are demanding the exact opposite.

In his 2008 book *The Way We'll Be*, pollster and trendwatcher John Zogby presented the results of a series of surveys his company, Zogby International, had done over the previous few years. After taking the pulse of the American people and their hopes and attitudes on everything from family life and job satisfaction to consumer preferences and political leanings, Zogby found many of his long-standing assumptions about the American dream thrown into question. For example, when his company undertook a comprehensive survey of consumer spending habits in 2005, he expected to find widespread confirmation of the stereotypical faults of his fellow citizens – "greed, overspending, obsession with luxury and brands, living beyond our means, failing to save for the future – the pursuit, in short, of an unreal reality."

Instead, stirring in the souls of Americans he discovered a "deep-felt need to reconnect with the truth of our lives and to disconnect from the illusions that everyone from advertisers to politicians tries to make us believe are real." People have grown tired of the spin-doctoring, marketing, and outright lies that emanate from political and corporate America. They are fed up with corporate frauds, such as Bernard Madoff's $50 billion Ponzi scheme or the scandals that put businessmen such as Bernard Ebbers (WorldCom), Kenneth Lay (Enron), and Conrad Black (Hollinger) behind bars. They have lost faith in a political culture that, on the heels of Bill Clinton's dissembling over the meaning of the word *is* (and his subsequent impeachment), could do no better as a replacement than the clueless George W. Bush, a man who signaled the start of a four-year insurgency in Iraq by declaring "mission accomplished" and whose reaction to the appalling federal response to the near destruction of New Orleans by Hurricane Katrina was to declare of the inept head of FEMA, Michael Brown, "Brownie, you're doing a heck of a job."

What Zogby found was a desire for authenticity.

But it is one thing to say you prefer the authentic. It is something else entirely to know what that means. Another extensive Zogby International survey, this one conducted in 2007, tried to determine what people mean when they talk about authenticity. When asked what character trait or factor made other people "authentic," more than one-third of respondents said "personality," while

38 per cent could do no better than to answer "other." When asked to pick from a list of words the one that offered the best definition of "authentic," 61 per cent chose "genuine" while 19 per cent opted for "real."

At best, these are mere synonyms that don't really clarify what is supposed to be so special about the authentic. But more worrisome is the fact that equating the authentic with the "genuine" or the "real" confuses things even further, since everything is, in a sense, genuine or real. After all, everything that exists just is what it genuinely is, and there is nothing that exists that is not real. Zogby is prepared to leave it at that: "Collectively, we Americans might not know exactly what 'authentic' is, but for the most part we know what it is not, and we know that we want whatever 'authentic' might be."

Two ideas come out of this. The first is that authenticity is a contrastive term, perhaps best understood negatively, by pointing to what it is not. The second is that whatever authenticity turns out to be, it is something people definitely want. That is, when something is described as "authentic," what is invariably meant is that it is a Good Thing. Authenticity is one of those motherhood words – like *community, family, natural,* and *organic* – that are only ever used in their positive sense, as terms of approbation, and that tend to be rhetorical trump cards. When a politician wants to plump for a preferred social policy, it is never a bad move to say that it will enhance communities or that it will benefit families. When companies want to sell a new product or service, calling it natural or organic is a sure-fire way of distinguishing it from its mass-produced competitors.

The upshot of all this is quite simple. In order to properly grasp what people mean when they talk about authenticity, we need to know the context in which it is used, and with what it is being contrasted. The authentic? Certainly. But as opposed to what?

The first chapter of the 2003 book *Authenticity: Brands, Fakes, Spin and the Lust for Real Life,* by British journalist David Boyle, is called "Living in an Artificial World." It begins with a neat narrative set-piece: the author sitting on the top of a hill with a small group of people waiting for a solar eclipse. As the moon trudges sullenly in front of the sun, Boyle notes how the world itself seems to have come to a stop: "The sheep lay down, the street lights in the nearest village in the distance across the moor came on automatically. It was a moment of almost spiritual doubt." Doubt, perhaps. But all in all a very *real* experience, the realness of it spoiled only by a man watching the BBC coverage of the eclipse on a portable television, afraid of what it would be like to experience life unmediated by technology. The crowd screams at the man to turn it off.

Many of us can sympathize with the crowd. It's getting to the point where you can't go to anything these days, be it a wedding or a funeral, a rock show or the opera, and not be bothered by someone tweeting away on a BlackBerry, chirping into a cell phone, or capturing the whole event for YouTubeal posterity on their camera phone or digital recorder. It is almost like we'd rather be any place other than where we actually, really, are. And if we have to be there, we'd prefer to see it mediated through technology.

For David Boyle, this preference for the virtual over the actual, the mediated over the real, is symptomatic of everything that is inauthentic about today's culture. Our world, he writes, is "dominated by spin-doctors, advertising, virtual goods and services. We are surrounded by the shoddy and the unreal, and by a global economy determined to foist it on us." Yet even as we find ourselves immersed in this world of the fast and the fake, Boyle thinks he discerns the early glimmer of a backlash, in which people are starting to demand an alternative to our virtual, spun, mediated, and marketed world. A revolution is stirring, and it is crystallizing around a single word: *authenticity*.

This is the same backlash against the culture that John Zogby discerned in his polling, but where Zogby's respondents had trouble articulating what it is precisely they are after, Boyle is starting to take us somewhere. He ends his first chapter with a list of ten "authenticity elements" he sees as essential to the backlash. Gathered under the notion of the authentic (or "real" – he treats them as equivalent terms), we find ideas such as the ethical, the natural, the honest, the simple, the unspun, the sustainable, the beautiful, the rooted, and the human.

In the various examples he gives to illustrate each element ("ethical" means things such as ethical investing or ethical corporate behavior; "natural" means anything not stained by technology, industry, chemicals, or mass production; "honest" means not marketed or branded), it is clear that "authenticity" involves the rejection of the various tributaries of mass society's current, including the media, marketing, fast food, party politics, the Internet, and – above all – the program of free markets and economic integration usually derided as "globalization."

This is a familiar critique. It is the same approach found in dozens of books published in the last few years, a genre established by Naomi Klein's *No Logo*, Todd Gitlin's *Media Unlimited*, Benjamin Barber's *Consumed*, and countless other works. These books are in turn little more than updates and elaborations of the "critique of mass society," a form of social criticism that has been in place since the 1960s and that arose as a direct response to the negative social effects of technology, bureaucracy, and consumer culture.

What makes Boyle's argument more interesting than the rest is that by gathering all these various critical movements under a single conceptual umbrella he calls "authenticity," he pushes the agenda in a direction that takes it beyond the familiar critique of conformity, consumerism, and capitalism. Unlike the critique of mass society, which is directed at an institutional structure and cultural imperative specific to the developed countries of the mid-twentieth century, the desire for the authentic is set against the backdrop of modernity itself, which encompasses the entire development of the West over the past two hundred and fifty years.

The seminal text on the rise of authenticity as a response to modernity is the 1972 book *Sincerity and Authenticity,* by literary critic Lionel Trilling, in which he explains that the term comes to us from the world of art history and museum studies. There, the question of authenticity refers to a work's history or provenance, so to ask whether a work of art is authentic is just to ask whether it is what it appears to be – a Ming vase or a Rothko painting. To affirm that a work is authentic is to say that it does, in fact, deserve the veneration or admiration that it appears to demand or that it was worth the price that was paid for it. Similarly, when we wonder whether an apparently ancient religious artifact is authentic, what we want to know is that it was actually used in ancient sacred rituals, as opposed to being manufactured for the sole purpose of selling it to tourists.

This distinction between how something *seems* and what it actually *is* – the distinction, that is, between appearance and reality – has been giving philosophers fits for centuries. As used in the earlier examples, it is what a philosopher would call an epistemological distinction because the determination of authenticity hinges on our knowledge of a fact of the matter. This looks like an authentic Ming vase. Is it? It certainly seems to, though it could be just a very competent forgery. Either way, there is a fact of the matter.

One of Trilling's great contributions was to show how this distinction, between how things appear to us and how they really are, eventually acquired a deeply moral dimension. This is captured by the familiar scene from *Hamlet*, as Laertes is preparing to head back to Paris. His father, Polonius, is following Laertes around, peppering his son with dull bits of fatherly counsel. But just as Laertes is ready to yell at his father, he is brought up short by these words:

This above all: to thine own self be true
And it doth follow, as the night the day,
Thou canst not then be false to any man.

Because Polonius is such a buffoon, Shakespeare scholars often write this off as just another bit of small-minded, selfish advice. But Trilling argues that this won't do because the sentence is simply too lucid, the sentiment too graceful. Polonius, he says, "has had a moment of self-transcendence," in which he has figured out that a loyalty to one's own true self is an essential condition of virtue. Trilling's reading of the passage is supported by the fact that the phrase "to thine own self be true" has passed into the language as a deft summary of an ideal that retains a powerful hold over our moral imaginations.

In its contemporary form, the moral dimension to authenticity amounts to the expectation that all the trappings of life, including your hopes and dreams, your job and family life, will reflect your true human purpose and potential. It is a profoundly individualistic ideal, understood as involving a personal quest or project that pushes self-fulfillment and self-discovery to the forefront of your concerns.

Authenticity has worked its way through our entire worldview, and our moral vocabulary is full of variations on the basic appearance-reality distinction, such as when we talk about someone exhibiting an "outer" control that masks an "inner" turmoil. When we meet people who are living inauthentic lives, we call them "shallow" or "superficial," as opposed to the more authentic folk who are "deep" or "profound." When it comes to relationships, authenticity is closely allied with the notion of sincerity, which demands a congruence between explicit ("outer") avowal and true ("inner") feeling. And it is because we are so concerned with this alignment of inner and outer selves that falseness, insincerity, and hypocrisy are seen as the great moral transgressions of our age.

The winter of 2005–06 provided a remarkable pair of examples. That fall, the term *truthiness,* coined by Stephen Colbert of *The Colbert Report* to mock the prevarications of government officials, became a part of the popular lexicon, while a panel of distinguished linguists at the American Dialect Society voted it the Word of the Year for 2005. At the same time, a small pamphlet called "On Bullshit," by philosopher Harry Frankfurt, became a surprising international best seller – considering that it was a twenty-year-old article originally published in an academic journal. But right from its opening sentence ("One of the most salient features of our culture is that there is so much bullshit"), the book's message obviously resonated with a public annoyed by a rash of corporate scandals and feeling wickedly deceived by the failure of the American and British forces to find weapons of mass destruction in Iraq.

Trilling suggests that the way authenticity "has become part of the moral slang of our day points to the peculiar nature of our fallen condition, our anxiety over the credibility of existence and of individual existences." What he is highlighting here is the biblical texture that permeates the whole discourse of authenticity: in the beginning, humans lived in a state of original authenticity, where all was harmony and unity. At some point there was a great discord, and we became separated from nature, from society, and even from ourselves. Ever since, we have been living in a fallen state, and our great spiritual project is to find our way back to that original and authentic unity.

What led to this apparent separation was nothing less than the birth of the modern world. Characterized by the rise of secularism, liberalism, and the market economy, modernity is the reason we have lost touch with whatever it is about human existence that is meaningful. Once upon a time religion, especially monotheistic religion, served as the objective and eternal standard of all that is good and true and valuable, and we built our society (indeed, our entire civilization) around the idea that living a meaningful life involved living up to that standard.

The search for authenticity is about the search for meaning in a world where all the traditional sources – religion and successor ideals such as aristocracy, community, and nationalism – have been dissolved in the acid of science, technology, capitalism, and liberal democracy. We are looking to replace the God concept with something more acceptable in a world that is not just disenchanted, but also socially flattened, cosmopolitan, individualistic, and egalitarian. It is a complicated and difficult search, one that leads people down a multitude of paths that include the worship of the creative and emotive powers of the self; the fetishization of our premodern past and its contemporary incarnation in exotic cultures; the search for increasingly obscure and rarified forms of consumption and experience; a preference for local forms of community and economic organization; and, most obviously, an almost violent hostility to the perceived shallowness of Western forms of consumption and entertainment.

The quasi-biblical jargon of authenticity, with its language of separation and distance, of lost unity, wholeness, and harmony, is so much a part of our moral shorthand that we don't always notice that we've slipped into what is essentially a religious way of thinking. The ease with which we talk about our alienation from nature, or the alienating nature of work, or the suburbs, or technology, is part of this language as well, hearkening back to our ongoing sense that we are fallen people.

The quest for authenticity is a quest to restore that lost unity. Where once we did it through actual religious rituals, prayer, and communion with God, now we make do with things such as Oprah's Book Club, which offers a thoroughly modern form of spirituality that is a fluid mix of pop-psychoanalysis, self-help, sentimentality, emotionalism, nostalgia, and yuppie consumerism. Or through our obsession with anything "organic" – organic beef, chicken, vegetables, cotton, dry cleaning, chocolate, or toilet paper. Similarly, a growing concern with a more local economy – local farmers, local bookstores, local energy production – reflects an underlying feeling that the holism of a small community is more valuable and more rewarding than the wasteful and messy free-for-all of mass consumerism. Finally, there is the almost visceral distaste for the market economy, driven by a conviction that the mere act of buying and selling is intrinsically alienating.

In all these guises, the search for the authentic is positioned as the most pressing quest of our age, satisfying at the same time the individual need for meaning and self-fulfillment and a progressive economic and political agenda that is sustainable, egalitarian, and environmentally friendly.

My central claim in this book is that authenticity is none of these things. Instead, I argue that the whole authenticity project that has occupied us moderns for the past two hundred and fifty years is a hoax. It has never delivered on its promise, and it never will. This is not because we aren't trying hard enough or are looking in the wrong places, or because the capitalists, politicians, and other purveyors of the fake are standing in our way. My argument is not that once upon a time we lived authentic lives – that we used to live in authentic communities and listen to authentic music and eat authentic food and participate in an authentic culture – and now that authenticity is gone. This is not a fairy tale.

Rather, the overarching theme of this book is that there really is no such thing as authenticity, not in the way it needs to exist for the widespread search to make sense. Authenticity is a way of talking about things in the world, a way of making judgments, staking claims, and expressing preferences about our relationships to one another, to the world, and to things. But those judgments, claims, or preferences don't pick out real properties in the world. There could never be an authenticity detector we could wave at something, like the security guards checking you over at the airport.

Lionel Trilling died in 1975, so he never lived to see cell phones and iPods, YouTube, the blogosphere, and Second Life. He missed out on the twenty-four-hour news cycle, $500 million presidential campaigns, and a built environment that wears advertising like a second skin. It is part of the ahistorical narcissism of our culture that we believe there is something extraordinary about the present. The pace of change is so rapid that we believe that our world is qualitatively different from what it was like even a few decades ago, and that our grandparents could not possibly understand the alienating pressures with which we are forced to cope. In some ways that is true. But part of this book's argument is to trace the origins of the authenticity quest, to show how, for all its apparent urgency and postmillennial relevance, very little has changed on the map since the battle lines were first drawn in the latter half of the eighteenth century.

Another part of this book's agenda is to argue that the problem was badly formulated from the start, and that the quest for authenticity has – at best – amounted to a centuries-long exercise in rainbow-chasing. More worrisome is the way our pursuit of the authentic ideal has become one of the most powerful causes of *in*authenticity in the modern world. To put it plainly: the contemporary struggle for genuine, authentic forms of living cannot be the solution to our problem, because it is the cause.

We are caught in the grip of an ideology about what it means to be an authentic self, to lead an authentic life, and to have authentic experiences. At its core is a form of individualism that privileges self-fulfillment and self-discovery, and while there is something clearly worthwhile in this, the dark side is the inherently antisocial, nonconformist, and competitive dimension to the quest. The hippie version of the authentic ideal, "doing your own thing," means standing out from the crowd, doing something other people are not doing. This creates competitive pressures to constantly run away from the masses and their conformist, homogenized lives. At the same time, when we take a closer look at many supposedly "authentic" activities, such as loft-living, ecotourism, or the slow-food movement, we find a disguised form of status-seeking, the principal effect of which is to generate resentment among others.

An even more severe consequence of the cult of authenticity can be seen in the urban – that is, black – culture of America's inner cities. In the jargon of the street, "keeping it real" ostensibly involves staying true to yourself, your family, and your community. In practice though, it amounts to a rejection of everything the Man wants you to do, like staying in school, working for a living, staying out of jail, and taking care of your kids. At best, "keeping it real" involves a high level of conspicuous consumption, spending a fortune on running shoes and clothes and jewelry; at worst, it serves as a justification for ignorance or even gangsterism.

In the public sphere, the desire for authenticity has contributed to a debased political culture dominated by negative advertising and character assassination. Meanwhile, a misguided nostalgia for illiberal and even premodern forms of political organization has fueled the forces of reaction, leading many otherwise well-meaning progressives to make common cause with dictators, fascists, and Islamic fundamentalists.

Many people are rightly concerned that our culture is locked into a competitive, self-absorbed, and hollow individualism, which gives us a shallow consumerist society completely lacking in genuine relationships and true community. But this leads to an uncomfortable paradox. After all, nobody ever admits to being shallow or false, and no one ever claims to love the artificial and the mass-produced. But if we all crave authenticity, how is it that the world seems to be getting more "unreal" every day? The argument of this book is that our misguided pursuit of the authentic only exacerbates the

—

ONE

THE MALAISE OF MODERNITY

IN ONE OF THE MOST FAMOUS ROAD TRIPS IN HISTORY, AN emaciated and notoriously untrustworthy Greek youth named Chaerophon trekked the 125 miles from Athens to the Temple of Apollo at Delphi to consult with the Oracle as to whether there was any man wiser than his friend Socrates. No, Chaerophon was told by the Oracle, there was no man wiser, and so he returned to Athens and informed Socrates of what the Oracle had said. At first Socrates was a bit skeptical, since it struck him that most of his fellow Athenians certainly acted as if they were wise about a great many things, while he, Socrates, didn't really know much about anything at all. But after wandering about the city for a while, questioning his fellow citizens about a range of topics (such as truth, beauty, piety, and justice), Socrates eventually decided that most of them were indeed as ignorant as he, but they didn't know it. He concluded that he was indeed the wisest Athenian, and that his wisdom consisted in the fact that he alone knew that he knew nothing.

Socrates should have known there was a bit of a trick to the Oracle's pronouncement. Inscribed in golden letters above the entrance to the ancient temple were the words *gnothi seauton* – "know thyself."

"Know thyself" thus became Socrates' fundamental rule of intellectual engagement, which he continued to deploy in Athens' public spaces, conversing at length with anyone who would indulge him and spending a great deal of downtime in the company of handsome young men. It cannot be said that his fellow Athenians appreciated this commitment to debate. In 399 BCE, Socrates was charged and tried for the crimes of teaching false gods, corrupting the youth, and "making the weaker argument the stronger" (a form of argumentative trickery called sophistry). At his trial, when the jury returned with a verdict of guilty on all counts, his accusers pressed for the death penalty. Given the chance to argue for an alternative punishment, Socrates started by goading the jury, going so far as to suggest that they reward him with free lunch for life. As for the other options – exile or imprisonment – he told the jury that he would not be able to keep his mouth shut on philosophical matters. Philosophy, he told them, is really the very best thing that a man can do, and life without this sort of examination is not worth living.

Socrates chose death rather than silence, and ever since he has been hailed for his integrity, a Christlike figure who was martyred for his refusal to sacrifice the ideals of intellectual independence, critical examination, and self-understanding. For many people, the Socratic injunction to "know thyself" forms the moral core of the Western intellectual tradition and its modern formulation – "to thine own self be true" – captures the fullness of our commitment to authenticity as a moral ideal.

For its part, the visit to the Oracle, with the cryptic pronouncement about Socrates having a special hidden characteristic, has become a stock motif of countless works of film and fiction, where the hero has to come to believe something about himself before he can help others. Perhaps the most hackneyed version of this is the scene in the film *The Matrix*, when Morpheus takes Neo to visit the Oracle. Morpheus believes that Neo is The One, the prophesied messiah destined to rescue humanity from the computer-generated dreamworld in which it has been enslaved. Neo is, understandably, a bit skeptical of his ability to serve as the savior of humanity. So Morpheus drags Neo off to see the Oracle, hoping that the good word from a maternal black woman who speaks in riddles while baking cookies will give Neo the boost of self-confidence he needs to get into the game and set about destroying the machines. Instead, the Oracle looks Neo in the eye and tells him he hasn't got what it takes to be the messiah. On his way out, she hands him a cookie and says, somewhat oddly, "Make a believer out of you yet." As Neo leaves, we see inscribed above the entrance to the kitchen the words *temet nosce*, which is Latin for "know thyself."

As it turns out, Neo is (of course) the messiah. The Oracle could not just come out and say so though, because Neo had to believe it himself. He had to buy into the whole worldview that Morpheus and his gang had laid out, about the rise of the machines, the scorching of the earth, and the enslavement of humanity. As Trinity tells Neo later on, it doesn't matter what Morpheus or even the Oracle believe, what matters is what Neo himself believes. The lesson is pretty clear. Before Neo can save humanity, he first has to believe in himself. The idea that self-knowledge and self-discovery are preconditions for social contribution is a thoroughly modern lesson, well steeped in the ethic of authenticity.

The Wachowski brothers were no doubt aware of the parallels they were drawing between Socrates and Neo (and, indirectly, between both of them and Jesus). Yet in *Sincerity and Authenticity*, Lionel Trilling makes it clear that this claim of continuity between the ancient world of Socrates and modern world of Oprah Winfrey and Eckhart Tolle is an anachronism, and that the authentic ideal is actually something relatively new. According to Trilling, the necessary element of authenticity – a distinction between an inner true self and a outer false self – only emerged in Western culture a few hundred years ago, toward the end of the eighteenth century. So despite superficial similarities, there is no real continuity between the Socratic dictum to "know thyself" and the thoroughly modern quest of self-discovery and self-understanding as an end in itself. What separates them is a yawning chasm between us moderns on the one side and the premodern world on the other.

What does it mean to be modern? That is a big and difficult question, and it has been the subject of a great many big and difficult books. One problem is that we often use *modern* as a synonym for *contemporary*, as when we marvel at modern technology or fret about modern love. Furthermore, even when we are careful to use *modern* to refer to a specific historical period, just what that is depends on the context. For example, historians sometimes refer to as "modern" the whole period of European history since the Middle Ages ended and the Renaissance began. Modern architecture, however, typically refers to a highly functional and unornamental building style that arose around the beginning of the twentieth century.

Here, I am concerned with modernity less as a specific historical epoch than as a worldview. To be modern is to be part of a culture that has a distinctive outlook or attitude, and while an important task for historians involves understanding why this worldview emerged where and when it did, it is essential to the concept of modernity that it is not tied to a particular place and moment. Modernity is what Marshall Berman, in his 1982 book *All That Is Solid Melts Into Air*, calls "a mode of vital experience – experience of space and time, of the self and others, of life's possibilities and perils – that is shared by men and women all over the world today." More than anything, modernity is a way of being, a stance we adopt toward the world and our place in it.

The rise of the modern worldview is marked by three major developments: the disenchantment of the world, the rise of liberal individualism, and the emergence of the market economy, also known as capitalism. Between 1500 and 1800, these three developments ushered in profound changes in people's attitudes toward everything from science, technology and art, to religion, politics, and personal identity. Put together, they gave rise to the idea of progress, which, as we shall see, does not necessarily mean "things are getting better all the time." More than anything, progress means constant change, something that many people find unpleasant and even alienating. But we're getting ahead of ourselves, so let's begin with the disenchantment of the world.

In the first season of the television series *Mad Men*, set in the advertising world of Madison Avenue in the early 1960s, gray-suit-and-Brylcreemed advertising executive Don Draper finds himself caught up in an affair with a bohemian proto-hippie named Madge. She drags Draper to parties and performance art clubs in Greenwich Village where he jousts with her anti-establishment friends over marketing and the moral culpability of capitalism. (Typical exchange: "How do you sleep at night?" "On a big pile of money.")

One night they end up back at an apartment, drinking and smoking pot and arguing once again. When one of the stoned beatniks informs Draper that television jingles don't set a man free, Draper replies by telling him to get a job and make something of himself. At this point, Madge's beatnik boyfriend chimes in with some classic countercultural paranoia: "You make the lie," he tells the ad man. "You invent want. But for them, not us." Draper has had enough, so he stands up, puts on his hat, and gives them some serious buzzkill: "I hate to break it to you, but there is no big lie. There is no system. The universe . . . is indifferent."

"Man," goes the extremely bummed reply. "Why'd you have to go and say that?"

If he'd bothered to stick around to continue the debate, Don Draper might have answered, *Because it is true.* For the most part, this exchange is nothing more than stereotyped bickering between hipsters and squares, of the sort that has been going on in dorm rooms and coffee shops for over half a century. But that last line of Draper's, about the indifferent universe, speaks to a deeper existential realization at the heart of the modern condition.

Once upon a time, humans experienced the world as a "cosmos," from a Greek word meaning "order" or "orderly arrangement." The order in this world operated on three levels. First, all of creation was itself one big cosmos, at the center of which was Earth. In fixed orbits around Earth revolved the moon, the sun, and the visible planets, and farther out still were the fixed stars. Second, life on Earth was a sort of enchanted garden, a living whole in which each being or element had its proper place. And finally, human society was itself properly ordered, with people naturally slotted (by un-chosen characteristics such as bloodline, birth order, gender, or skin color) into predetermined castes, classes, or social roles.

Whatever else it might have been, this was a place of meaning, value, and purpose, with each part getting its identity from knowing its place in the whole and performing its proper function within an organic unity. For both the ancient Greeks and, later, medieval thinkers, this fundamental order could be described by the notion of the "great chain of being," a strict hierarchy of per-fection stretching from the rocks and minerals, up through the plants and animals, to humans, angels, and God. In this geocentric and homocentric cosmos, humanity found itself trebly at home, comfortably nestled like a Russian doll within a series of hierar-chies. Earth was the most important part of creation, and humans were the most important beings on Earth. Finally, human society was itself a "cosmos," a functional and hierarchical system – of slaves, peasants, commoners, tradespeople, nobles, and so on – in which each person's identity was entirely determined by their place in that structure.

This was the worldview in which Socrates (or at least, the man who is revealed through Plato's writings) operated. For Socrates, self-discovery involved little more than coming to understand where you fit in the grand scheme of things. On this reading, the oracular injunction to "know thyself" would be better expressed as "know thy place." This is not to say that people did not have ambitions, emotions, or deeply felt desires, just that these were not important to helping you discover your place in the world.

Our films and fiction are full of romantic stories about the sons of medieval cobblers who fall in love with, woo, and win the local nobleman's daughter, or Georgian scullery girls who find themselves swept off their feet and up to the castle by the young prince who is sick of dating fake society girls, but these are nothing more than projections of our own assumptions and values onto a world in which such behavior was literally inconceivable. The work of Jane Austen is so important precisely because it marks the transition from that world to a more modern sensibility – most of her stories hinge on her characters' nascent individualism straining against the given roles of the old social order.

Almost every society that has ever existed has seen its world as "enchanted" in one way or another, from the polytheism of the ancient Greeks and Romans to the strict social roles of Chinese Confucianism to the form many of us are most familiar with, the monotheistic religions of Judaism, Christianity, and Islam. What is characteristic of traditions of this sort is that they are what we can call "comprehensive doctrines," in that they purport to explain and justify a great deal about life on Earth, how the world works, and why human society is structured as it is, all within a common metaphysical framework.

Consider Catholicism, which at its peak was a powerful comprehensive doctrine that began by providing an explanation for the origins of the universe (God made it in seven days) and life on Earth (God made Adam out of dust, and Eve out of one of Adam's ribs). Additionally, it provided a moral code (the Ten Commandments), along with a justification (God commands it), backed up by a sanction for violations (you'll burn in hell). Finally, it explained the meaning of life, which consists of spiritual salvation through communion with God, mediated by the priesthood. Science, politics, morality, spiritual succor – the Catholic church is a one-stop explanatory shop, serving needs existential, political, social, and scientific.

What a comprehensive religious tradition does is ensure that everything that happens on Earth and in human society makes sense. In the end, everything happens for a reason, as it must be interpreted in light of what God wills, or what He commands. This is a version of what philosophers call teleological explanation – explanation in light of ultimate purposes or goals (from the Greek *telos*, meaning "end.") The disenchantment of the world occurs when appeals to ultimate ends or purposes or roles being built into the very fabric of the universe come to be seen as illegitimate or nonsensical.

The big steamroller of Christianity was science, as a series of discoveries – from Copernican heliocentrism to Darwinian natural selection – played an important role in shaking up humankind's sense of its place in the scheme of things. But even though science has progressively discredited any number of specific religious claims, there is no necessary antagonism between science and religion at the deepest level, and for many scientists, scientific inquiry is just a way of coming to understand the mind of God. As the great Renaissance polymath Sir Francis Bacon put it, "a little science estranges a man from God; a lot of science brings him back."

So scientific discoveries alone are not enough to kick us out of the enchanted garden, and over the centuries religion, especially the Catholic Church, has shown itself to be very adept at accommodating the truths of divine revelation to those discovered through scientific inquiry. One standard move is to avoid resting the truth of the faith on any particular scientific fact and simply assert that however the world is, it is like that because God intended it that way, while another is to "protect" the divine origins of the human soul by simply declaring it off-limits to empirical inquiry.

Pope John Paul II employed both of these methods in 1996, when he conceded that evolution was "more than a hypothesis." Darwin might be right about evolution through natural selection, the Pope said, "but the experience of metaphysical knowledge, of self-awareness and self-reflection, of moral conscience, freedom, or again of aesthetic and religious experience, falls within the competence of philosophical analysis and reflection, while theology brings out its ultimate meaning according to the Creator's plans."

This is what that biologist Stephen Jay Gould called NOMA, short for non-overlapping magisteria. NOMA is a sort of explanatory federalism, where the scope of human experience is carved into watertight compartments in which each type of explanation – scientific, religious, aesthetic, metaphysical – is sovereign within its own competency. If we accept the NOMA gambit, then disenchantment cannot be the inevitable outcome of a string of increasingly devastating scientific discoveries. As the Pope's address makes clear, religion can never be chased completely off the field as long as people are willing to accept the validity of certain noncompeting forms of explanation, even if that amounts to little more than accepting "God wills it" as an account of why some things happen. For true disenchantment to occur, the scientific method of inquiry must be accepted as the *only* legitimate form of explanation.

If your high-school experience was anything like mine, the phrase *scientific method* conjures up memories of science labs with Bunsen burners, test tubes, and wash stations. We were taught a rigid formula that supposedly captured the essence of scientific inquiry, consisting of a statement of the problem, a hypothesis, a method of experiment, data, and a conclusion. (Problem: Do the same types of mold grow on all types of bread? Hypothesis: Yes. Method of Experiment: Leave lunch in locker every day for a month.) But the scientific habit of mind is more abstract than this, and it actually has little to do with experimentation, data collection, and analysis. At the core of scientific thinking are two elements.

First, the commitment to explanation in terms of general laws and principles. Second, the recognition of the open-ended and ultimately inconclusive character of scientific progress.

Back when people believed that gods roamed and ruled Earth, the explanation of natural phenomena was basically a soap opera. Why was last night's thunderstorm so intense? Because Hera caught Zeus cheating on her again, and when she started throwing crockery he replied with thunderbolts. Why were all our ships lost at sea? Because the captains forgot to sacrifice a bull to Poseidon before they left port, and he showed his displeasure by whipping up a storm with his trident.

It doesn't take a great amount of insight to appreciate that these little stories don't actually explain anything at all. They are ad hoc accounts invented after the fact as a way of imposing some sort of order on a world that is often totally unpredictable. Worse, they give us no way of predicting the future. When will Zeus cheat again? Who knows. Will sacrifice to Poseidon keep our ships safe? Probably not, since he appears to get angry for the most arbitrary reasons, and plenty of captains who did remember the sacrifice have had their ships founder on a lee shore. Clearly, if we're looking for some way of both understanding the past and giving some reasonable guidance to what will happen in the future, we need something better.

The person usually credited as the first to turn away from supernatural soap opera as a way of making sense the world was an Ionian Greek named Thales of Miletus, who lived sometime between 620 and 546 BCE. Thales was interested in the nature of matter and how it is able to take the tremendously diverse forms that make up the furniture of the world, and he proposed that the primary organizing principle of all things is water. We don't have a proper record of why Thales believed this, since most of what we know of his thought comes to us second-hand from Aristotle.

According to Aristotle, Thales may have got the idea by observing the role of water in creating and preserving life: "That the nurture of all creatures is moist, and that warmth itself is generated from moisture and lives by it; and that from which all things come to be is their first principle. . ." As a theory of everything, "all is water" is not much of an advancement, but what gives Thales his well-deserved reputation as the first true philosopher is a conceptual innovation we can call the *generality of reason*.

Thales realized that stories about Zeus cheating on Hera or Poseidon being irritated by the lack of sacrifices don't help us understand why they behaved the way they did – these stories are pseudo-explanations that give only the illusion of providing understanding. Real understanding requires some sort of formula that shows us how events of this type can be expected to have effects of that type, both looking back (as an explanation of what happened) and forward (as a prediction of what will happen). For this, we need to understand events in the light of laws or general properties of some sort.

Once we have the idea of the generality of reason, we are armed with a tremendously powerful cognitive tool, since the notion that the world operates according to predictable general laws is what gives us logic, science, and technology, as well as the principles of impartiality and equality in the ethical and moral realms. And so by suggesting that water was the originating principle of all things, Thales took a fateful step away from the random and usually bizarre plot twists of the supernatural soap opera and took a stab at explaining the world in terms of general principles that would enable us to understand why things are the way they are as well as predict how they might be in the future.

Yet this, too, is still a long way from the disenchantment of the world. The world can be a rational, ordered place, explicable in terms of general laws and principles, and still be a cosmos. Thales showed us how to leave soap opera behind, but there is still plenty of room for recourse to arguments underwritten by final explanations in terms of God's will, plan, or intentions. What drives teleology from the field once and for all is the realization that science is a progressive endeavor that can *in principle* never come to an end.

It was the late nineteenth-century German sociologist Max Weber who showed us how the disenchantment of the world occurs once, and only once, we fully appreciate the relentlessly progressive nature of science. As he argued, scientific discoveries are by their very nature meant to be improved upon and superseded. We can never say we have achieved the final truth, because there is always the possibility of further explanation in terms of broader or deeper laws. For Weber, the commitment to science

> means that principally there are no mysterious incalculable forces that come into play, but rather that one can, in principle, master all things by calculation. This means that the world is disenchanted. One need no longer have recourse to magical means in order to master or implore the spirits, as did the savage, for whom such mysterious powers existed. Technical means and calculations perform the service.

The term is overused, but Weber's final repudiation of magical thinking represents a genuine *paradigm shift* in our outlook on the world, and is a giant step toward becoming fully modern. The most significant consequence of this is that the disenchanted world is no longer a cosmos but is now a universe. It is Don Draper's morally indifferent realm, consisting of mere stuff – energy and matter in motion – that neither knows nor cares about humans and their worries. The philosophical significance of this is that the world can no longer serve as a source of meaning or value. That is, no statement of how the world is can, by itself, validate any conclusions about what ought to be the case. We are no longer entitled, for example, to the argument that just because some groups are slaves, slavery is their natural *and therefore justified* condition.

This gap between reasoning about what is and reasoning about what ought to be, and the illegitimacy of proceeding from the first to the second, was famously noticed by David Hume. Known as "Hume's guillotine," it cautions philosophers to be on their guard against unwarranted inferences.

Along with everything else, disenchantment transformed our understanding of the self, it privileged a utilitarian philosophy that saw the maximizing of happiness as the ultimate goal, and it encouraged an instrumental and exploitative approach to nature through the use of technology. A key effect of disenchantment, though, was its action as a social solvent, helping break up the old bonds in which individuals and groups found their place within larger class-based divisions or hierarchies. When people no longer have a proper place in the scheme of things – because there is no scheme of things – they are, in a sense, set free. They are free to make their own way, find their own path.

Thus, the disenchantment of the world leads directly to the second major characteristic of modernity, the rise of the individual as the relevant unit of political concern.

One effect of disenchantment is that pre-existing social relations come to be recognized not as being ordained by the structure of the cosmos, but as human constructs – the product of historical contingencies, evolved power relations, and raw injustices and discriminations. This realization wreaked a great deal of havoc with long-established forms of social organization, and it came at an auspicious time. At the end of the eighteenth century, even as people were coming to appreciate the conventional and arbitrary origins of the traditional hierarchies in which they found themselves embedded, the nascent industrialization of parts of Europe was pushing them into greater proximity, thanks to large-scale migrations from the country into the cities and town centers.

This double contraction of distance (of social class and of geography) inspired a root-and-branch rethink of most elements of political authority. For the first time, people began asking themselves questions such as: Who should rule? On what basis? Over whom should power be exercised, and what are its scope and limits? What inspired these sorts of questions was the relative decline of groups, castes, nations, and other collectivities, and the emergence of the individual as the central unit of political concern.

A recurring theme in the literature on the birth of modernity is the idea that political individualism was itself the result of the prior rise of religious individualism in the sixteenth century. The decisive event here is the Protestant Reformation, which Tennyson called "the dawning of a new age; for after the era of priestly domination comes the era of the freedom of the individual." Unlike Catholics, whose faith is mediated through the supreme authority of the Church, Protestants regarded each person's interpretation of the Bible as authoritative. The Protestant relationship to God is personal and unmediated, with piety residing not in good acts or confession of sins, but in the purity of will. As Martin Luther put it, "it is not by works but by faith alone that man is saved." By abolishing the need for a separate caste of priests, Protestantism effectively turned religion into a private matter, between the individual and God. Furthermore, the emphasis on salvation through faith fueled a psychologically inward turn, in which the examination of one's conscience took center stage.

The Reformation was indeed an important step in the emergence of political individualism. In particular, it fed a stream of modernity that influenced the rise of authenticity as a moral ideal. Yet the religious individualism of the Reformation was itself enabled by a much more profound development, the emergence of the centralized state. The deep connection between the rise of the modern state and the emergence of the individual is not always fully appreciated. They are actually just two aspects of the same process, and it is no coincidence that the individual became the focus of political concern just as the centralized state was beginning to consolidate its power in the sixteenth century.

The state is such a distinctively modern institution that most of us find it difficult to imagine any other way of carving up the world, so much so that we habitually describe territories that employ other forms of government – such as Afghanistan or Sudan – as "failed states." It was not always so. There are many ways of governing a territory, and the state is only one of them. As it emerged out of the Middle Ages, the state vied with a handful of other, sometimes overlapping, forms of political organization, including those based on kinship, tribal affiliation, feudal ties, religion, and loose confederal arrangements such as the Holy Roman Empire. With the idea of the state comes the notion of *sovereignty*, which in your standard Politics 101 textbook is usually defined as something like "the monopoly over the use of force in a territory" or "the exercise of supreme legislative, executive, and judicial authority" over a well-defined geographical area. The paired ideas here are *supremacy* and *territoriality*; together, they embody the form of government we know as the sovereign state. Instead of governing along with or indirectly through local tribesmen or warlords, members of the nobility, various estates, or the Church, the sovereign asserts his unlimited and unrestricted authority over each and every member of society.

What does this have to do with political individualism? After all, what does it matter to a serf, to a woman, or to a Jew whether they are oppressed by the whims of a local lord, husband, or cleric, or by the decrees of a distant king? Eliminating the middle man may be a useful step toward more efficient tyranny, but it is hard to see how it involves a decisive concession to the principle of individual liberty. Nevertheless, that was the precise effect of the rise of the modern state. As Oxford political scientist Larry Siedentop puts it, the modern state is a Trojan horse, carrying with it an implicit promise of equality before the law:

> The very idea of the state involves equal subjection to a supreme law-making authority or power – the sovereign. To speak of a "state" is to assume an underlying equality of status for its subjects . . . once there is a sovereign agency or state, inherited practices no longer have the status of law unless they are sanctioned by the sovereign.

And so the emergence of the state was the political sidecar to the process of disenchantment that was already underway. Just as disenchantment stripped away the metaphysical foundations that justified inherited social classes as fixed roles, the state cut through the various overlapping layers of political authority that had accumulated over the centuries like so many coats of paint by establishing a direct and unmediated relationship between the sovereign and the individual subject. In contrast with every other form of political organization, the state is first and foremost a collection of individuals. They may have any number of social roles – tradesperson, wife, baron, priest – but these are merely secondary add-ons. Individuality is now the primary social role, shared equally by all.

This distinction between primary and secondary social roles is only possible within a sovereign state. To see this, try to imagine a collection of historical types that includes an ancient Egyptian slave, a medieval serf, an Untouchable from India in the eighteenth century, and a nineteenth-century Frenchman. Now imagine you could go back in time and ask each one of them to describe the underlying bedrock of their identity, perhaps by simply asking each of them, "What is your status?" All the first three could say in response is that they are, respectively, a slave, a serf, and an Untouchable. Their identity is entirely constituted by their inherited social roles. In contrast, the Frenchman could reply that, while he occupies a number of secondary social roles (such as courtier, landowner, husband), he is first and foremost a free-standing *citoyen*, an individual whose formal equality with all other Frenchmen is guaranteed by the very nature of the French republic.

A second, related distinction is between *law* and *custom*. In a society where people are born into a social role – such as an Egyptian slave or a Confucian wife – in which they have an obligation to obey another in virtue of that social role, all social rules have the same status. In Confucianism, for example, there is no distinction between temporal laws and religious decrees – both carry equal weight because there is no essential difference between the principles that govern the Kingdom of Heaven and those that govern the Kingdom of Man. Similarly, in a theocratic state there is no essential difference between religious decrees and temporal laws, and in a feudal system there is no difference between the expressed preferences of the lord and his actual decrees.

But when there is an established state claiming a monopoly over political authority, we see the arrival of a distinction between *laws*, which are the explicit and obligatory commands of the sovereign backed up by a sanction, and *customs*, which might be enforced through social pressure but which have no legitimate legal backing. Of course, some systems retain a certain hybrid character, and a state such as India remains imperfectly sovereign thanks to the ongoing influence of the caste system, despite the fact that it has been outlawed. But that's the entire point: the claim about the tight relationship between individualism and the state is not that once a state is formed, people no longer try to contest the state's legitimacy by enforcing religious decrees or assuming certain caste privileges, it is that these attempts are illegitimate by the state's own self-understanding. Note that much of what we call nation-building (which is actually state-building) in a place such as Afghanistan involves trying to subordinate the influence of local warlords or religious leaders to that of the central government in Kabul.

The two distinctions, between primary and secondary social roles and between positive law and mere custom, are creatures of that quintessential institution of modernity, the sovereign state. And once they are in place, they are able to evolve into their fully developed form, the liberal distinction between the public realm, which is within the law's reach, and the private realm, which is a sphere of personal conscience, worship, choice, and pursuit. But for this to emerge, the concept of sovereignty has to be drained of its unlimited reach, through the notion of the limited state.

Probably the best essay on the tremendous gap between social roles under feudal forms of government and in the modern state is the famous scene in *Monty Python and the Holy Grail* where King Arthur approaches a peasant digging in the dirt to ask for directions. Arthur begins by introducing himself as "King of the Britons," a claim greeted with a wink of incredulity. Arthur persists in asking the peasant for directions, but the peasant protests that Arthur has not shown him proper respect. After Arthur replies that he is, after all, the king, an argument ensues over the nature of legitimate government and the scope of the state's writ, which ends with the peasant insisting that supreme executive authority derives "from a mandate from the masses, not from some farcical aquatic ceremony" – a reference to the Lady of the Lake who gave Arthur the sword Excalibur and anointed him King.

The scene has grown stale over the years, thanks largely to bores who insist on quoting it verbatim at parties (complete with accents), but what made it work in the first place is the cognitive dissonance involved in a medieval peasant rattling on to his king about a modern concept such as responsible government. It might seem obvious to us moderns that a ruler the people didn't vote for has no legitimacy, but the Python scene reminds us what an aberration that notion is, considered against the sweep of human history.

Virtually all premodern societies, including those within the Christian tradition, are built around what philosophers call a "perfectionist" value system. They conceive of society first and foremost as a community, organized around a shared conception of human perfection, of what promotes human flourishing and what matters to the good life. The main role of the established authority is to promote the moral and spiritual perfection of the people. It does this by using political power to exercise moral authority and to declare which beliefs and behaviors are virtues and which are vices, condoning the former and condemning the latter. The full coercive apparatus of the state is used to enforce these sanctions.

The perfectionist attitude toward the state was captured by the traditional French saying *Une foi, un loi, un roi* (one faith, one law, one king), and it was only after the devastating French Wars of Religion in the sixteenth century that Europeans started to come around to the idea that the continent's divisions were probably here to stay. The basket of virtues we take to be characteristic of Western individualism began with religious toleration, and it was something not so much chosen as forced upon an economically shattered and morally exhausted Europe as a second-best alternative to perfectionism.

It was through a similar process that the liberal commitment to individual rights arose, through widespread revulsion in the face of the horrors that could be visited upon the population by an increasingly powerful and centralized state. Sure, monarchs had always fought one another, marching out in the spring, fighting a few skirmishes on a suitable battlefield, then marching home in the fall, but the people were largely left out of things. Only when the state began to turn its power of coercion against its own citizens – that is, when dictatorship of absolute monarchy evolved into the tyranny of the Inquisition – did people start to consider that maybe the business of the state shouldn't be the moral perfection of the people, and maybe its authority ought to be limited in certain important ways.

This shift in thinking is marked by the transition in thought between two seventeenth-century philosophers, Thomas Hobbes and John Locke. Hobbes was quite certain that the citizens of a commonwealth would prefer an absolute sovereign to the nasty and brutish condition of the state of nature (the "war of all against all"), but Locke – writing in the aftermath of the Glorious Revolution – finds this ridiculous. He proposed that the state be divided into separate branches, where the citizens have a right to appeal to one branch against another. This is the beginning of *constitutionalism*, or the idea of the limited state. The main principles of constitutionalism are that the state is governed according to the rule of law; that everyone is equally subject to the law; and that the scope of what is a legitimate law is limited by a charter of individual rights and liberties. A constitutional state is one that foregrounds its respect for the autonomy of the individual and his or her free exercise of reason, choice, and responsibility.

To a large extent, respect for individual autonomy follows directly from the rejection of natural hierarchies and assigned social roles. When people possess formal equality under the state, and when they no longer owe any natural and enforceable obligations to one another, then a great many decisions – about how to worship, what to do for fun, whom and how to love – become a matter of personal choice. As the future prime minister of Canada, Pierre Elliott Trudeau put it in 1967 when, as a young Justice Minister, he introduced a bill decriminalizing homosexual acts between consenting adults: "There's no place for the state in the bedrooms of the nation." But this sort of autonomy is valuable only to the extent that the decisions people actually make are respected and protected. It would be a hollow form of autonomy indeed if the state told people that they were free to make various choices but then turned around and allowed them to be persecuted for those decisions.

That is why any suitably robust commitment to autonomy must be accompanied by a firm commitment to preserving the legal distinction between the public and the private realm. Just where the line is to be drawn will differ from one state to the next, but a liberal society must carve out some minimal space for the protection of individual conscience and the pursuit of private goals. This puts the question of individual rights onto the agenda: as philosopher Ronald Dworkin has argued, "Rights are best understood as trumps over some background justification for political decisions that states a goal for the community as a whole." That is, a right is a trump card that preserves a sphere of private individual action against decisions taken by the state in the name of the common good. Locke summarized things with the declaration that everyone had the right to life, liberty, and property, the ultimate consequence of which was a ground-up rethink of the appropriate relationship between the state, morality, and the good life.

An essential part of this system of individual liberty that emerges from the Hobbes-to-Locke trajectory is a species of economic individualism, also known as a market economy, also known, to its critics anyway, as *capitalism*. The emergence of the global market economy marks the third step in the development of the modern world, a step that was to prove the most transformative yet also the most controversial and socially disruptive.

Mention capitalism in the context of the last years of the eighteenth century and what immediately clouds the mind are proto-Dickensian images of urban squalor, belching factories, and overworked and threadbare six-year-olds being beaten by rapacious landlords in top hats. While this image is not completely inaccurate, the use of the term *capitalism* puts a misleading emphasis on material forces, while neglecting the powerful ideals motivating this new economic individualism. In particular, focusing on material relations (and even the class struggle) obscures the role of

individual autonomy, the rise of the private sphere, and the importance of contract in conceiving a fundamentally new approach to
the morality of economic production and consumption.

On the economic side of things, the most important consequence of Locke's liberalism was the idea that the public good
could be served by individuals pursuing their private interest. With
the "privatization of virtue," ostensible vices such as greed, lust,
ambition, and vanity were held to be morally praiseworthy as long
as their consequences were socially beneficial. In his *Fable of the
Bees* (subtitled *Private Vices, Publick Benefits*), Dutch-born physician
Bernard de Mandeville argued that luxury, pride, and vanity were
beneficial because they stimulated enterprise. Mandeville was an
obvious influence on Adam Smith's *Wealth of Nations*, particularly
on Smith's infamous metaphor of the invisible hand of the market
that describes a positive, yet unintended, consequence of self-
interested behavior. Smith argued that each individual, seeking
only his own gain, "is led by an invisible hand to promote an end
which was no part of his intention," that end being the public interest. "It is not from the benevolence of the butcher, or the baker, that
we expect our dinner," Smith wrote, "but from regard to their own
self interest."

This is clearly at odds with almost all popular moralities, including Christianity, which emphasize the importance to public order
and to the common good of individual sentiments of benevolence
and public-mindedness. But it is no great leap from Locke's economic individualism to the idea that what matters to morality are not
intentions, but outcomes. What does it matter why people behave
the way they do, as long as society is better off for it? If we can
harness the pursuit of private interest to positive social outcomes, it
is hard to see what anyone could have to complain about. The theory
that best served this new morality was utilitarianism, summarized
by philosopher and social reformer Jeremy Bentham as the principle of the greatest good for the greatest number.

What ultimately validated the utilitarian pursuit of happiness – that is, hedonism – as a morally acceptable end in itself was the first great consumer revolution, which began in the second half of the eighteenth century, in which both leisure and consumption ceased to be purely aristocratic indulgences. Not only did consumerism become accessible to the middle classes, it became an acceptable pursuit; buying stuff, and even buying into the spiritual promise of goods, came to be seen as a virtue. Precisely what spurred the dramatic increase in consumption in the 1770s and 1780s is a matter of considerable dispute in the sociological literature, but everyone agrees that the crucial development was the emergence of fashion consciousness amongst the masses. Sure, there had always been fashion trends of some sort, insofar as what was considered appropriate style, material, and color changed over time. But this was almost always the result of social emulation amongst the aristocracy, and shifts happened slowly, sometimes over the course of decades. But in the 1770s, the "fashion craze" made its appearance, characterized by the distinctly modern feature of rapid shifts in accepted popular taste. In 1776, for example, the "in" color in the streets of London was something called *couleur de noisette*; a year later, everyone who was anyone was wearing dove gray.

In order for there to be a consumer revolution, there had to be a corresponding revolution in production. That is because consumption and production are just two ways of looking at the same unit of economic activity; one person's consumption is another person's production. That is why any significant change on the demand side of the economy must be accompanied by an equal shift on the production side. And indeed it was, via the great convulsion we call the Industrial Revolution. Historians argue over the precise dating of the Industrial Revolution, but most agree that it began with a number of mechanical innovations in the British textile industry in the 1760s. These innovations quickly spread to France, Holland, and the rest of Western Europe, and across the Atlantic to Canada and the United States, along with similar and almost simultaneous gains in other industries, especially iron-smelting, mining, and transportation.

The Industrial Revolution affected almost every aspect of the economy, but there were two main aspects to the growth in innovation: first, the substitution of work done by machines for skilled human labor, and second, the replacement of work done by unskilled humans and animals with inanimate sources of work, especially steampower. Put these together – machines replacing men, machines replacing animals – and you get a tireless supply of energy driving increasingly sophisticated machines. This marked the death of the cottage industry and the birth of the factory, where power, machines, and relatively unskilled workers were brought together under common managerial supervision.

As David Landes points out in his book *The Wealth and Poverty of Nations*, what made eighteenth-century industrial progress so revolutionary was that it was contagious. "Innovation was catching," writes Landes, "because the principles that underlay a given technique could take many forms, find many uses. If one could bore cannon, one could bore the cylinders of steam engines. If one could print fabrics by means of cylinders . . . one could also print wallpaper that way."

The same idea held for printing, textile manufacturing, tool machining, and countless other industrial processes, where a discovery made in one area reinforced those made in others, driving innovation along an ever-widening front. As ever-higher incomes chased ever-cheaper consumer goods, the economy became locked into a virtuous cycle of growth and development, with each new innovation serving as a stepping stone to another. After centuries of relative stagnation, for the first time in the history of the world people had to get used to the fact that the future would not resemble the past, that there would be *progress*. That is, people had become modern.

When it comes to the Industrial Revolution, the two long-standing questions are: Why Britain? Why then? These are fantastically difficult questions and are virtually impossible to distinguish from the larger question of why some countries are rich and some are poor. Is it a matter of race or climate? Culture or environment? Religion or politics? You can take your pick of answers, depending on your taste and intellectual temperament. Regardless, it is almost certainly no accident that the Industrial Revolution happened where and when it did.

If you were given the opportunity to design a society optimized for a "growth and innovation" economy, there are a number of characteristics you would be sure to include. You would want it to be a society of intellectual freethinkers, in which people had the right to challenge conventional thinking without having to worry about persecution or censorship. This freedom of thought would include a commitment to the virtues of empiricism and the willingness to accept the reports of inquiry and observation at face value and to pursue these inquiries wherever they may lead. Your ideal society would have a legal mechanism for translating the fruits of research and invention into commercial enterprise by respecting property rights and the freedom of contract. And there would be few social or legal taboos against consumption, allowing for the flourishing of a market for new goods, in the form of a mass consumer society. Freedom, science, individualism, and consumerism: that's a difficult agenda for any society, one that few countries in the world today have managed to get their heads around. But as a first draft of modernity, it is not a bad description of late-Enlightenment Britain.

When we think about the deployment of power, technology, and human resources, what usually comes to mind are architectural marvels such as pyramids or cathedrals, or grand public enterprises like the Manhattan Project or the space program. That is, we think of public works along the lines of Kublai Khan's Xanadu, stately pleasure domes brought into existence by the decrees of grand rulers or the ambitions of great states. But the most remarkable aspect of the Industrial Revolution is that it was powered almost entirely by the private, household consumption desires of the middle classes. In their pursuit of personal happiness and self-fulfillment through economic development and consumption, the British nation of shoppers and shopkeepers unleashed a force unlike anything the world had ever seen.

No one understood this better than Karl Marx. *The Communist Manifesto* begins with what seems to be an extended and genuine appreciation of the bourgeoisie, a group that has managed to conquer the modern state and, in contrast with the "slothful indolence" of the middle ages, has unleashed the active forces that lie within. It has done this by hitching the latent power of technology to the forces of private competition, and in so doing has remade the world.

> The bourgeoisie, during its rule of scarce one hundred years, has created more massive and more colossal productive forces than have all preceding generations together. Subjection of nature's forces to man, machinery, application of chemistry to industry and agriculture, steam navigation, railways, electric telegraphs, clearing of whole continents for cultivation, canalization or rivers, whole populations conjured out of the ground – what earlier century had even a presentiment that such productive forces slumbered in the lap of social labor?

Even Marx concedes that this was not a wholly bad occurrence. Once it had the upper hand, the bourgeoisie "put an end to all feudal, patriarchal, idyllic relations." It shrugged off all the old, arbitrary ties that bound men to one another in virtue of who their ancestors were or what color their skin was. The bourgeoisie cleared a social and political space for the emergence of new visions of the good life, new possibilities for what people can do and who they might become.

Yet capitalism proved to be a universal solvent, eating away at the social bonds between people in a given society as well as the cultural barriers that formerly served to separate one society from another. In place of the family or feudal ties, of religiosity, of codes of conduct like chivalry and honor, there is now nothing left but the pitiless demands of the cash nexus, the rest having been drowned "in the icy water of egotistical calculation." Meanwhile, all that is local and particular succumbs before the relentless cosmopolitanism and consumerism of the world market. National industries falter before the relentless scouring of the globe by industry for resources and manpower, while every nation's cultural heritage – its poetry, literature, and science – are turned into homogenized commodities. In sum: read Marx on capitalism and you realize how little was added to his original critique by the critics of globalization of the late twentieth century.

Capitalism is able to do this by exploiting the almost limitless capacity that humans have for enduring perpetual upheaval and change, in both their personal and public affairs. A capitalist society puts tremendous pressure on people to constantly innovate and upgrade, to keep on their toes. They must be willing to move anywhere and do anything, and "anyone who does not actively change on his own will become a passive victim of changes draconically imposed by those who dominate the market." For the bourgeoisie, life comes to resemble the world of the Red Queen in *Through the Looking-Glass*, where it takes all of the running one can do just to stay in the same place.

Modernity, then, is the offspring of three interlocking and mutually supporting developments: the disenchantment of the world and the rise of science; the emergence of political individualism and a taste for liberal freedoms; and the technology-driven gale of creative economic destruction known as capitalism. These gave us a new kind of society and, inevitably, a new kind of person, one who has learned to thrive in a milieu in which freedom is equated with progress, and where progress is nothing more than constant competition, mobility, renewal, and change. In what is probably the most succinct and evocative one-paragraph description of modernity you will find anywhere, a passage that perfectly captures the almost delirious light-headedness of the modern worldview, Marx writes,

> Constant revolutionizing of production, uninterrupted disturbance of all social conditions, everlasting uncertainty and agitation distinguish the bourgeois epoch from all earlier ones. All fixed, fast frozen relations, with their train of ancient and venerable prejudices and opinions, are swept away, all new-formed ones become antiquated before they can ossify. All that is solid melts into air, all that is holy is profaned, and man is at last compelled to face with sober senses his real condition of life and his relations with his kind.

But no sooner had people learned to enjoy the fruits of modernity (all that stuff, all that freedom) than they started complaining of the bitter pit in the center; even as some were tallying up the gains, others were keeping track of the inevitable losses.

And losses there certainly were. First, each person lost his or her given place in society, through the breakdown of the old order, the destruction of pre-existing hierarchies, and the demise of ancient ways of living. Second, humanity as a whole lost its place in the grand scheme of things. The world was no longer an ordered cosmos but instead a chaotic universe, a cold and indifferent realm of mere matter in motion.

Not everyone finds this loss of meaning acceptable, just as not everyone is happy with the related demise of nature as a source of intrinsic value. As for individualism, freedom is all well and good, but is it worth the price? Is it worth it if the result is a narrowing of vision, a shallowness of concern, a narcissistic emphasis on the self? The consumer culture that arose thanks to the privatization of the good life may be good for the economy, but is it good for the soul? For a great many people, the answer to questions of this sort was – and is – a clear "no." As they see it, what is lacking in modernity is an appreciation for the heroic dimension of life, higher purpose having been sacrificed to the pursuit of what Alexis de Tocqueville called "small and vulgar pleasures." As for the pleasant-sounding principle of "economic individualism," it quickly showed its stripes as the hard realities of laissez-faire capitalism hit home. Adam Smith's vaunted division of labor may have made for great efficiency gains at the pin factory, but it didn't make working there any more enjoyable. If anything, work became rigid, mechanical, and boring, and somewhere between the cottage and the factory workers stopped identifying themselves with the products of their labor. They became what today we call wage slaves, trading their effort for money, but not necessarily for meaning.

And so the victories of modernity look from another perspective like defeats, and purported gains appear as losses. The main problem is not that modernity has eliminated unwanted hierarchies and stripped away unwarranted privileges. It is that it has dissolved every social relation, drained the magic from every halo. We have replaced the injustices of fixed social relations with the consumer-driven obsession with status and the esteem of others, and where we once saw intrinsic meaning and value we now find only the nihilism of market exchange. Critics have found it useful to gather all of these problems and objections to modernity under the term *alienation*.

If you have been paying any attention at all to the world around us, you will have noticed that there seems to be a lot of alienation going around. Husbands are alienated from their wives, students are alienated from their teachers, voters are alienated from their politicians, and patients are alienated from their doctors. Everyone thinks the mass media are alienating, especially thanks to all the advertising. Religious people find the permissiveness of our secular society alienating, and some believe that alienation is what motivates terrorists. People who live downtown find the suburbs alienating, while suburbanites feel the same way about life in the big anonymous city. The world of work seems to have tapped a rich vein of alienation, if the success of comics such as *Dilbert* and television shows such as *The Office* are any indication of popular sentiment. In modern society, we are all alienated from nature and from one another, although perhaps that is only because ultimately we are all alienated from ourselves.

If modernity has a lot of explaining to do, the concept of alienation is doing most of the explaining. It is a substantial burden to lay on a single word, particularly one that is routinely left undefined. For many people, alienation is like victimhood: if you feel alienated, then you are, which is why the term commonly serves as a synonym for *bored, powerless, ignorant, unhappy, disgusted, aimless*, and just about every other negative descriptor for one aspect of our lives or another.

To say that something is "alien" just means that it is in some sense different, foreign, or "other," hence the use of terms such as *alien life form* or *illegal alien*. To alienate something is to make it foreign or separate, and someone is alienated when they feel disconnected or detached from their friends, their families, or their jobs. Trade or commerce is a form of alienation: property is alienated when ownership is transferred from one person to another. Abstract concepts can also be alienated; social contract theorists talk of citizens alienating (that is, transferring) certain of their natural rights to the state. In its strictest sense then, alienation is just a separation or lack of unity between pairs of people, groups, and things.

But the dictionary can only take us so far, since what matters is not what the word means, but the various uses to which it is put.

When social critics talk about alienation, they tend to veer between talking about alienation as a psychological or as a social phenomenon. They may often be related, but they are actually logically distinct. Psychological alienation refers to your attitude, emotions, or feelings toward whatever situation you find yourself in, whether it is your work, or your marriage, or your living environment, and it typically manifests itself as dissatisfaction, resentment, unhappiness, or depression.

Social alienation, on the other hand, is not concerned with whether you are unhappy or resentful. Instead, it looks at the social, political, or economic structures and institutions in which people find themselves embedded. The discordant effects of social alienation are caused by a "lack of fit" between people's actual behavior and the norms or expectations of their environment, and they manifest themselves in various ways. For example, a high crime rate in an inner-city neighborhood might be attributed to the fact that the youth want to skateboard in the church parking lot but the cops keep chasing them away. Thus, these youth might find themselves alienated from their urban environment. Similarly, a high absentee rate amongst employees in a large corporation might be the result of an alienating corporate culture that squeezes unique individuals into cookie-cutter cubicles that are hostile to free, creative thought.

Regardless of which type of alienation we are talking about, it is vital to keep in mind the following: *Just because you are alienated, it does not mean that there is a problem and that something ought to be done about it.* Both of the types of alienation, the psychological and the social, are just descriptions of certain states of affairs. The first describes someone's state of mind, the second is an account of certain social relationships among persons, groups, and institutions.

Consider again the world of work. The contemporary office, with the long lines of cubicles (what novelist Douglas Coupland coined as "veal fattening pens") manned by cut-and-paste worker drones who are simmering with barely restrained rage, is a thoroughly worked-over metaphor for contemporary alienation. The anonymous nature of bureaucratic organizations, and the routine and mechanical nature of the work, seems completely at odds with any reasonable understanding of what might make someone happy. Anyone who doesn't feel alienated in such an environment, we think, must be either drugged, insane, or lobotomized. But so what? Nobody says you are supposed to like your job, and nothing says it is supposed to be fulfilling. To put it bluntly: there's a reason why they call it work, and there is a reason why they pay you.

This line of reasoning should be ringing a few bells. It is nothing more than a restatement of Hume's guillotine, which forbids us from drawing any moral conclusions from strictly empirical or descriptive premises. This is-ought problem might seem strange in the context of a discussion about alienation. Nobody refers to an institution as alienating unless they are expressing moral or political disapproval, and to describe someone as alienated is to make it very clear that you think something needs to be done about it. In fact, for many of its proponents, the great virtue of thinking about society in terms of alienation is that it provides a way of doing an end-run around Hume's guillotine and bridging the is-ought gap.

To see how this is supposed to work, consider the way we think about disease. When a doctor diagnoses you with an illness, you might say that she is simply *describing* a state of affairs. If you have cancer, it means that you have uncontrolled cell division occurring in some tissues of your body. To say you have malaria is to say that a certain protozoan parasite is multiplying within your red blood cells. But of course we want to say more than that. Illness isn't just a description of the state of the body like any other; the difference between being healthy and being sick is not the same as the difference between having brown or blond hair, or between standing up or lying down. Yes, sometimes we would like to stand, sometimes we prefer to lie down, but that preference depends on other desires and purposes we have at the time. There is nothing intrinsically wrong with standing up or with lying down. In contrast, the very idea of what it means to be sick (and lurking in the etymology of the word *dis-ease*) is that something is not right, that the body has been disturbed from a normal or natural state to which it ought to return or be restored. In its everyday usage, disease is more than a description of how things are – it implies a way things *ought* to be.

Alienation theory tries to bridge the is-ought gap by treating alienation like a disease: it not only describes a state of affairs, it also considers that state of affairs as abnormal or unnatural. It carries with it an implicit normative judgment, a preference for a natural, nonalienated state that ought to be restored. To do this, alienation theory needs something to play the role of analogue to health. Just as medicine has an account of what constitutes normal or natural health, alienation theory needs an account of what constitutes normal human life. It needs a theory of human nature or of self-fulfillment that is not just relative to a given place or culture, or relative to an individual's desires at a certain time. It needs an account of human flourishing that is in some sense natural or essential. If the alienation we moderns feel is in fact a malaise, a sort of illness, then what is needed is an account of what it would mean to end the discord, to restore the absent sense of coherence or unity. That is, for a theory of alienation to do any work, it needs a corresponding theory of authenticity.

This, in a nutshell, was the burden of Romanticism. The Romantic response to modernity was an attempt to transcend or mitigate the alienating effects of the modern world and recoup what is good and valuable in human life. The key figure here is French philosopher Jean-Jacques Rousseau.

TWO

A FALSE RETURN

JEAN-JACQUES ROUSSEAU DIDN'T HAVE A GREAT START TO LIFE. OR perhaps it is more accurate to say that his mother didn't have a great start to his life, since she died during childbirth, leaving Rousseau's father devastated. But he had a happy enough childhood, reading voraciously, and he was raised by his father to believe that Geneva was a city as magnificent as ancient Rome. When Rousseau was ten, though, his father was forced to leave town under threat of imprisonment (he was accused of drawing his sword against a captain of the militia), so young Jean-Jacques was sent to the nearby town of Bossey to live with a Protestant minister, Monsieur Lambercier (also named Jean-Jacques), and his sister. Rousseau spent two happy years with the Lamberciers – happy, that is, except for one distasteful episode with a comb. In retrospect, Rousseau thought, that was the moment when everything started to go downhill.

One morning, young Jean-Jacques was sitting in a room off the kitchen doing his homework. A servant came in and left a number of combs belonging to Mademoiselle Lambercier on the stove to dry. When the servant returned for the combs, she found that one of them had all its teeth broken down one side. As no one else had been in the kitchen, suspicion naturally turned to Rousseau. As he told the story:

I was questioned and denied having touched the comb. M. and Mlle Lambercier jointly lectured, pressed, and threatened me, but I steadfastly maintained my denial. They were convinced otherwise, however, and so sure of themselves that they swept aside all my protests, even though this was the first time that I had been suspected of an outright lie. They took the matter seriously, as it deserved. The mischief, the lie, and my stubborn refusal to confess all seemed worthy of punishment.

The whole sordid affair left a powerful impression on Rousseau, and from it he drew two lessons. First, he came to appreciate how the distinction between appearance and reality can be a source of injustice. By all appearances he was guilty of having destroyed the comb, and none of his accusers had the power or clairvoyance to drill down through the world of *seems* to the world of *is*, to see the truth that lay in Rousseau's heart (not to mention the recent past). "I stuck to my own view, and all I felt was the cruelty of an appalling punishment of a crime I had not committed."

More importantly, the mystery of the broken comb taught Rousseau the value of deceit. After all, if appearances are the only things that matter, then why should we ever concern ourselves with anything else? What is the point of trying to be open and transparent when you risk being systematically misread and misunderstood? The distinction between how things *seem* and how they actually *are*, which had up until then been a fairly abstract notion for the young intellect, was revealed as a source of deep loneliness and separation, and Rousseau concluded that in a world where people are judged entirely by appearances, the only sensible thing to do is to wear a mask.

Jean-Jacques Rousseau was far from the first philosopher to recognize that appearances can be deceiving. The centerpiece of Plato's *Republic* is his famous allegory of the cave, in which Plato says that people are like prisoners chained in a cave and facing a wall, unable to turn their heads. Behind them burns a fire, and between the fire and the prisoners there is a parapet, along which puppeteers walk, holding up objects – such as a vase, or a rabbit – that cast shadows on the wall of the cave. As Plato imagines it, the prisoners mistake the objects in the shadow-puppet world for the real thing, unable to see the true entities that dance on the parapet behind them.

Plato uses this little story as a way of explaining the nature of the world and our knowledge of the things in it. His argument is that just as the vases and rabbits the prisoners are pointing to are only shadows of the real entities behind them, the objects we pick out in the world, that we point to and name, are in fact shadow-objects of the higher reality, in which resides what he calls the "Form" of those objects. For Plato, this higher, truer reality, the realm of the Forms, is something we can only grasp conceptually in our minds.

In the seventeenth century, René Descartes turned the appearance-reality distinction from a metaphysical problem into a full-blown crisis of knowledge. Observant fellow that he was, Descartes noted that our senses frequently let us down. We hear voices in the hall, but step out only to find no one there; we chase a friend down the street and are embarrassed when we finally catch up with a stranger. Even the things that seem quite certain, like the fact that I am sitting here typing at a computer, are open to doubt. After all, Descartes notes, has it not happened that I have on occasion had a dream so clear and vivid that it seemed real?

Once you start down this skeptical path, it is hard to know where to stop. Descartes followed it right to the end, wondering whether it was possible that he was wrong about absolutely everything. How do I know, he wondered, that I am not being deceived at every turn by an evil genius or demon, who "has employed his whole energies in deceiving me?" In such a case, Descartes assumes that everywhere he steps the demon has laid a trap for his credulity, and he is forced to "consider myself as having no hands, no eyes, no flesh, no blood, nor any senses, yet falsely believing myself to possess all these things."

What brings Descartes back from the brink where sheer paranoia tips over into outright solipsism is his famous assertion that there is one thing he knows for sure, one truth that the evil genius can't fool him about, and that is his – Descartes' – existence as a thinking being: "I am, I exist is necessarily true each time that I pronounce it, or that I mentally conceive it."

I think, says Descartes. Therefore, I am.

Indeed, by the time Descartes was done with it, the "appearances can be deceiving" line had become a bit of a philosophical cliché. But then Rousseau kicked new life into the old workhorse by hitching the appearance-reality distinction to an equally venerable Fall-of-Man parable, and doing so changed the terms of debate in Western thought.

Before the incident with the comb, Rousseau lived in an Edenic state of unity and innocence. He was at one with his own mind, transparent to himself, and consequently felt a sense of unmediated connection, respect, and intimacy with the people in his life. But the accusation and his subsequent unfair conviction led to a double separation. Because he is no longer trusted, he has become emotionally separated from those he felt closest to. At the same time, he now sees that all that matters is appearance, and he learns the value of insincerity, secrecy, and lies. He is alienated from others, but also now from his own true self.

What Rousseau came to realize is that the gap between appearance and reality is not just metaphysical (as Plato thought) or epistemological (as it was for Descartes) but that it has a moral dimension as well, since it is the source of all that is wrong with the world. Appearance is the realm of guilt, reality the domain of innocence. Goodness is found in the realm of *is*, while *seems* is the root of all evil. The appearance-reality distinction is the original sin that introduces the knowledge of the evil of separation into his world, a break that is all the more cruel because it involves a crime that Jean-Jacques swears he did not commit.

This is a Fall-of-Man story with a twist. In the original biblical version, Adam and Eve are pretty much the agents of their own expulsion from the Garden. But what is unique in Rousseau's account is that no one is to blame. As even he admits, the comb situation did not look good and the Lamberciers were fully justified in blaming him. The problem ultimately lies not with men and their bad intentions, but with society and the inevitable friction it introduces into relations between people. Society is necessarily the land of appearances, and it is society that introduces evil into the world, in the form of the quest for prestige, status, wealth, and esteem.

So how do we restore the lost goodness, the unity before the fall? For Rousseau, as for many others who have followed him, there are two possible answers. We need to fix society or, if it cannot be fixed, we must turn our backs on it.

For a man convinced that status-seeking, selfishness, and insincerity are the great evils of civilization, Jean-Jacques Rousseau was a stupendously vain and egotistical man. When he was sixteen, he fled Geneva and made his way to France, where he was taken in by Madame de Warens, a twenty-nine-year-old widow. He lived with her for most of the next eight years, and while he initially called her Mommy, they became lovers when Jean-Jacques was twenty-two.

Restless and ambitious, Rousseau seems to have spent most of the next decade working on various get-famous-quick schemes. He invented a new form of musical notation, wrote a few plays, and then, in 1743, managed to secure himself a position in Venice as an assistant to the French ambassador. It wasn't quite what he expected. Instead of being given the respect and deference he felt was his due as a diplomat, Rousseau was treated like the servant he was. He fought with the ambassador, returned to Paris, and moved in with a house servant named Thérèse Lavasseur, who would be his companion for the rest of his life. It was a bizarre relationship. She was dull and dimwitted, and they eventually sent all five of their children off to a state-run home for foundlings. Rousseau offered a number of excuses for this appalling behavior, pleading both poverty and the superior ability of the state to teach the children a worthwhile trade.

While in Venice, Rousseau had determined that music was his ticket to glory. He wrote an opera called *Les Muses Galantes* (The Amorous Muses), which he gave to the famous composer Rameau for comments. Rameau told him that it was quite bad, a criticism that the notoriously thin-skinned Rousseau actually took to heart. Nevertheless, he kept plugging away at his music career until 1749, when he caught wind of an essay competition established by the Academy of Dijon, inviting entries addressing the question "Has the revival of the Sciences and Arts contributed to improving morality?"

In a burst of contrarian insight, Rousseau decided to answer the question in the negative, presenting a short, sharp indictment of Enlightenment values and the modern world. This *Discourse on the Arts and Sciences* claimed that, far from improving and edifying mankind, the growth of the arts and sciences have made people soft and conformist, sandpapering away the coarse edges of our natural individuality. Obsessed with the demands of public decorum and politeness, we have become restrained and insincere.

The essay was a sensation, winning first prize in the competition. It also established Jean-Jacques Rousseau as a first-rank political theorist and gained him the celebrity for which he had been searching his entire adult life. But there was more to come. Four years after the publication of the first *Discourse*, he wrote his masterpiece, the *Discourse on Inequality*. That essay restated Rousseau's basic case against modernity, extending and amplifying the themes of the first *Discourse*: society is characterized by status competition, selfishness, and the pursuit of private interest, which causes men to be alienated from one another and from themselves. The result is hypocrisy and insincerity, a thoroughly inauthentic existence in which appearances reign supreme.

The second *Discourse* is a just-so story, a speculative account of the historical development of the human race from the most primitive beginnings, through early tribal society, into agricultural communities, and then modern society. It begins with a description of the "state of nature," a conceit common to many political theorists of the time. The state of nature is supposed to describe the "natural" condition of mankind – that is, how he would live if there were no government, no civil society, no laws, and no state to assert its monopoly over the use of force.

The best-known account of the state of nature comes to us from Hobbes, in his political treatise *Leviathan*. The popular perception of Hobbes is that he was a pessimist, arguing that man is fundamentally bad and that the state of nature would quickly degenerate into "the war of all against all." This is not quite right. What Hobbes did believe is that each person was a *psychological egoist*, that is, that each person is always motivated by self-interest. Furthermore, Hobbes believed that in the absence of a state or other coercive authority, each of us has what he described as the "natural right" to take an action they feel necessary to preserve and promote their freedom, security, and well-being. In the state of nature, you have the right to come up while I'm sleeping, bonk me on the head, and take my food. Likewise, I have the right to do the same to you.

For Hobbes, the state of nature is a large, multiplayer prisoner's dilemma, where what is good for everyone, collectively, is undermined by each person's individual rational calculations. Without a coercive authority to enforce cooperation, each of us retreats into tactics of self-preservation that are collectively self-defeating. It is not human nature, but the structural lack of restrictions on people's behavior, that led Hobbes to assert, infamously, that life in the state of nature would be "solitary, poor, nasty, brutish, and short."

Rousseau has a rather different account of the physical condition of man in the state of nature. As he imagines it, it is a rather congenial sort of place, in which man enjoys a life of isolation, equilibrium, and self-sufficiency:

> I see an animal less strong than some, and less agile than others, but, upon the whole, the most advantageously organized of any: I see him satisfying his hunger under an oak, and his thirst at the first brook; I see him laying himself down to sleep at the foot of the same tree that afforded him his meal; and there all his wants are completely supplied.

Congenial then, but not exactly a paradise. Man still has to fend for himself – against the wind and the rain and predators, not to mention the daily need to satisfy hunger and thirst and fend off exhaustion. Nature is a drill sergeant, and those who manage to survive and thrive are rewarded with a strong and robust constitution. Consequently, illness is rare, and the only real health problems are those that arise during infancy or from old age. All of the modern infirmities, from gout and bad digestion to depression and sleeplessness, are unheard of. Life is simple, uniform, and solitary, but it is neither nasty, nor brutish, nor short. There is no war of all against all because there is no reason for it. Nature provides all that a man needs to be solitary and self-sufficient.

Rousseau also disagrees with Hobbes in his account of man's moral condition in the state of nature. In contrast with Hobbes's monotonic "psychological egoism" (the claim that we are utterly self-interested), Rousseau sees human nature as characterized by two basic drives, the first of which he calls self-love (*amour de soi*) and the second, pity: "One of them interests us deeply in our own preservation and welfare, the other inspires us with a natural aversion to seeing any other being, but especially any being like ourselves, suffer or perish."

A few things are worth noting here. First, Rousseau sees self-love as motivated by nothing more than the need to promote the survival and flourishing of the individual, by satisfying each individual's rather modest needs. Self-love is about finding food and shelter, little else, but even this minimal amount of self-interest is moderated by the second drive, pity. "It is this pity which hurries us without reflection to the assistance of those we see in distress; it is this pity which, in a state of nature, takes the place of laws, manners, virtue." The upshot is that nature endows man with a moral psychology highly conducive to peace and order, captured by the maxim "Do good to yourself with as little prejudice as you can to others."

So what went wrong? How did we get from the congenial state of nature to the cutthroat selfishness of modern life?

Initially, it looks as if Rousseau is going to place the blame squarely on the usual suspect, private property. The second part of the *Discourse on Inequality* begins by blaming centuries of crime, war, murder, and other horrors on the first person who, "after enclosing a piece of ground, took it into his head to say, *this is mine*." But it turns out that the invention of property isn't the primary culprit – it is itself merely a downstream effect of changes in human social organization that began much earlier. It is crucial to Rousseau's argument that there is no one to blame, that there is nothing anyone did, no essential flaw in man's character. The problems of modernity emerge as the more or less inevitable consequences of social interaction.

Rousseau has already conceded that nature doesn't coddle her children – not all fruits are low-hanging, not all obstacles are easily surmounted. But because they are possessed of an innate ingenuity and resourcefulness, humans gradually found ways of dealing with the various problems nature presents them. They learned to make tools like the hook and line for fishing, and the bow and arrow for hunting. They discovered how to build shelters, to use animal fur to keep warm, to tame fire and use it for cooking, and various other ways of making the world a more accommodating place.

But the principal effect of all this ingenuity (we certainly can't call it progress) was to accentuate the natural inequalities between individuals. These differences in strength, speed, courage, and intelligence don't mean much in the solitary and simple state of nature. But as men became more industrious, they got better at controlling and manipulating their environment. They were able to build huts, then maybe a cabin, which they then felt obliged to defend. This was the beginning of property. It was also the beginning of the nuclear family, for the first time making it possible for parents and their children to live together as single unit, a "little society" held together by conjugal and parental love.

This changed everything, says Rousseau. Before, social intercourse was transient and fleeting. But then these transient relationships started to congeal into a more settled form of life, with people living permanently near one another and coming into regular contact. And, inevitably, this nascent society led to the idea of comparison, or what Rousseau calls "relations." It becomes obvious to everyone that some men are stronger, or smarter, or more attractive, or more courageous, than others. In becoming aware of how they compare with others, men got into the habit of self-regard, and "thus it was that the first look he gave into himself produced the first emotion of pride in him."

For Rousseau, pride does not come before the fall, pride *is* the fall. The sense of pride that comes from comparing yourself to your fellow man, and coming out ahead, gives birth to a new motivation, and new form of self-love, which Rousseau calls *amour-propre*. Unlike the natural and useful feeling of self-love he calls *amour de soi* (which is just a drive for self-preservation tempered by pity), *amour-propre* is essentially other-regarding. It is nothing less than the quest for status, from which all the evils of civilization follow.

In the first place, *amour-propre* leads to selfishness. Beyond the comfort and leisure having lots of stuff affords, men soon found that the accumulation of property could be a source of both status and domination, and so the pursuit of wealth became a means of acquiring prestige over others. But the problem with commerce is not just that it enables the endless acquisition of wealth. Commerce is itself an intrinsically alienating form of social interaction because it takes the direct and natural relation of mutual esteem and replaces it with relationships mediated by stuff. Because commercial transactions are motivated entirely by the desire for private gain, human contact becomes thoroughly instrumentalized. We treat others as means to our own selfish ends, not as ends in themselves. Men are quickly alienated from one another, as human contact becomes nothing more than an excuse for exploiting others.

It would be bad enough if the quest for property, wealth, and status served only to alienate men from one another. But as Rousseau argues, the real problem with society is not social alienation, but self-alienation. Once *amour-propre* comes to dominate the relations between men, everyone becomes obsessed with appearances and with questions such as who sings or dances the best, who is the best-looking, or the strongest, wittiest, or most eloquent. Status becomes the only good worth pursuing:

> Men no sooner began to set a value upon each other, and know what esteem was, then each laid claim to it, and it was no longer safe for any man to refuse it to another. Hence the first duties of politeness, even among savages; and hence every voluntary injury became an affront, as besides the hurt which resulted from it as an injury, the offended party was sure to find in it a contempt for his person often more intolerable than the hurt itself.

In such a world, deception becomes a necessary survival skill. In a society dictated by relations of vanity and contempt on the part of social superiors, and of the envy and shame of inferiors, it becomes imperative to appear better than you actually are. The mediated world of *seems* is now paramount, and the unmediated and unmasked world of *is* ceases to matter.

The target of Rousseau's scorn was, first and foremost, Paris and Parisians. He hated the place – it rendered him "alienated, frustrated, embittered, and desperate," as one scholar has put it. But the broader target of his indictment was civilization itself. He just saw Paris as the most proximate example of all that was wrong with modernity. It is important to realize that Rousseau didn't deny that modernity had its attractions, and he was perfectly willing to admit that the quest for honor had spinoff benefits in the form of great achievements in the arts and sciences. But when it came to weighing costs and benefits, he felt certain that what we had gained from civilization was far outstripped by what we had lost, in the way of economic self-sufficiency, psychological transparency, and connection to nature.

When it comes to coping with the downside of the modern world, there are two lines of approach. We can try to eliminate the causes of our problems or, alternatively, we can work toward mitigating the effects. That is, we can see about changing society and eliminating competition and inequality or we can focus on building stronger, more self-sufficient individuals within the sphere of modern life. As it turned out, Rousseau thought the second approach had the best chance of success, although he is probably most widely known for trying the first tack, because of a misinterpretation of his account of the state of nature.

After Rousseau had sent him a copy of his second *Discourse*, the leading lamp of the Enlightenment, Voltaire, responded with a sardonic little note:

I have received, sir, your new book against the human species, and I thank you for it . . . no one has ever been so witty as you are in trying to turn us into brutes: to read your book makes one long to go on all fours. Since, however, it is now some sixty years since I gave up the practice, I feel that it is unfortunately impossible for me to resume it: I leave this natural habit to those more fit for it than you and I.

With that, Voltaire endorsed what has become the dominant popular conception of Rousseau's views on human society, namely, that he was a romantic primitivist full of nostalgia and longing for mankind's precivilized past. Today, Rousseau is commonly credited (or, more often, blamed) for introducing the concept of the noble savage into our moral vocabulary, although he himself never used the term.

In fact, the term *noble savage* was first used by poet John Dryden in 1672, after which it pretty much disappeared from the language until it was resurrected, almost two hundred years later, by a British anthropologist named John Crawfurd, who pinned its conceptual origins on Rousseau. Crawfurd's scholarly ambition was to demonstrate that all other societies were inferior to Europeans, and he figured he could use a straw version of Rousseau's argument as a foil.

Jean-Jacques Rousseau has been the piñata of anthropology ever since. In his essay "Enter the Noble Savage," anthropologist Roger Sandall argues that, sure, Dryden may have coined the term, but "nothing much happened until Rousseau picked up the idea and ran with it." Thanks to Rousseau, "the tribal world was morally transformed and the savage himself had been redeemed," and primitivism is no longer regarded with the horror that it deserves.

There is no question that, whether he talked explicitly of noble savages or not, Rousseau's rather dismal account of civilization (and his correspondingly enthusiastic account of the state of nature) had considerable uptake among his contemporaries. Characteristic of the neo-Rousseauian genre is the work of a French Benedictine monk named Dom Deschamps, who dreamt of a world free of the petty jealousies and enviousness that arose out of prideful men competing with one another in a market economy. In a passage that makes groups like the Khmer Rouge and the Taliban seem urbane in comparison, Deschamps proposed a world where intellectuals would be banned and everyone would live together in a hut, "work together at simple tasks, eat vegetarian food together, and sleep together in one big bed of straw. No books, no writing, no art: all that would be burned."

While his specific thesis – no art, no metal, no meat – might seem a bit extreme, in his general themes Deschamps was articulating a position that has attracted a great deal of sympathy from political leftists of a certain stripe over the past two hundred and fifty years. Modern civilization is alienating, while primitive societies promise a return to our lost unity and natural wholeness, where we can avoid the status competition and raw commercialization of society and embed ourselves in a true community based on simple, nonexploitative relationships. In this view, the search for our lost authenticity is essentially an exercise in retrieval, as we hearken back to our own premodern past.

As both history and anthropology, Rousseau's idea of the state of nature is total bunk. If contemporary evidence is anything to go by, there is nothing peaceful, congenial, or even terribly solitary about tribal life. Instead, it is a world of "despotic chiefs, absurd beliefs, revolting cruelty, appalling poverty, horrifying diseases, and homicidal religious fanaticism" (a state of affairs which has been almost completely eradicated from the modern world). The percentage of people killed in battle in hunter-gatherer societies such as the Dani of Papua New Guinea, or the Yanomamo of South America, is about 30 per cent, which is more or less the same rate found among chimpanzees. As one anthropologist writes of tribal warfare in New Guinea, "This last place on Earth to have remained unaffected by modern society was not the most peaceful but one of the most warlike ever encountered."

And it appears to have always been this way. Far from an aberration or occasional activity motivated by desperation or fear, warfare between primitive societies "was incessant, merciless, and conducted with the general purpose of annihilating the opponent." About two-thirds of prehistoric tribes were in a state of near-constant warfare, with casualty rates that, had they been suffered by populations in the twentieth century, would have added up to about two billion deaths.

In short, the romantic thought that primitive societies lived in peace and security, in a harmonious, transparent, and authentic relationship with the earth and with one another, appears to be not consistent with the facts. Perhaps nothing captures the absurdity of it all better than the *New Yorker* cartoon that shows two cavemen sitting cross-legged on the ground facing each other across a fire. One says to the other: "Something's just not right – our air is clean, our water is pure, we all get plenty of exercise, everything we eat is organic and free-range, and yet nobody lives past thirty."

As a criticism of Rousseau's theory of civilization, this sort of historical fact-checking kind of misses the point. Rousseau was not an anthropologist, and his goal was not to turn Paris into some sort of Papua-sur-Seine. The fact is, he never claimed that the "state of nature" as he described it had ever existed, and even if it had, he didn't think it was a condition to which we could (or even should) return. Rousseau's critique of civilization was certainly an exercise in philosophical nostalgia, but if anything it was a nostalgia for Europe's own lost innocence.

For Roger Sandall, this is a fateful concession, and he accuses Rousseau of operating in what he calls the "hypocritical mode" of reasoning. On the one hand, Rousseau repeatedly concedes that we could never return to the state of nature, since it probably never existed in the first place. But at the same time, the whole tone of his critique of civilization implies that a tribal state would be far preferable to the predatory character of modern society, and the underlying theme is that civilization is a false and nasty cloak that we need to shake off. And so Rousseau's writing, argues Sandall, "creates a pervasive atmosphere of ambiguous make-believe and insincerity."

If true, this would be a devastating irony – the master theorist of transparency, relentless critic of falsity, tireless proponent of authenticity, found guilty of insincerity and hypocrisy. But that is too quick. Rousseau was essentially a social critic, and he saw his main role as pointing out the power relationships in which people find themselves embedded through no choice or fault of their own, with a special emphasis on the way the rich and the powerful exploit the poor and the weak. That, more than anything else, is the force behind his famous quotation "Man is born free but everywhere he is in chains."

Despite the way his most hostile critics have tried to paint him, Rousseau wasn't offering a facile primitivist line about the natural and best state of man being to prance around on a mountain side like a solitary billy goat. Rather, he was pointing to the far more immediate vices of Parisian society, such as the fact that the whole game was rigged in favor of an aristocracy that had not earned, and did not deserve, its status, and suggesting that there might be a better way of organizing things. So a more charitable reading of Rousseau is to think of his state of nature as a "regulative ideal" that is unattainable in practice but that can be used to evaluate actual social institutions and relationships and to measure our progress toward a more egalitarian and less exploitative society.

If there is a major failure in Rousseau's writing, it is not that he was an inept historical anthropologist, nor that he was just another hypocritical moralist. The real problem with Rousseau is that he was so utterly disgusted by the particular failures of Parisian society, and so wracked by his own insecurity, that he was unable to appreciate anything that such a society had to offer. While he certainly knew what he didn't like, there was little sense of balance or understanding of the tradeoffs involved in civilization, and so what someone like Sandall sees as hypocrisy might be better interpreted as something closer to antimodern tunnel vision. In that sense, on the living tree of Rousseau's intellectual descendants, there is one group that has enthusiastically adopted this tunnel vision and developed it into a root-and-branch condemnation of the modern world.

December 21, 2012, is a date you should probably mark on your calendar. It is the winter solstice, and astronomers tell us that it is also the day when the sun will be perfectly lined up with the center of the Milky Way galaxy for the first time in 26,000 years – though it isn't clear why that is significant. More importantly though, it marks the end of a 5,126-year cycle on the ancient Mayan "long count" calendar. And the significance of *that* is that it means the world is going to come to an end, thanks to some sort of astronomical Y2K. The exact mechanism of the apocalypse is unknown, but if you troll around the Internet you can find any number of speculative scenarios. Most of them presume that there'll be a sort of massive ecological collapse and extinction event caused by a combination of global warming, deforestation, peak oil production, overfishing, overpopulation, suburbia, megacities, bird flu, swine flu, consumer electronics, hedge funds, credit default swaps, and fast food.

Let us call the people who seriously foresee the coming apocalypse "declinists," and their animating philosophy "declinism." What motivates declinism is an attitude so pessimistic that it is almost theological: not only are things worse than they used to be, but they're getting worse with every passing year. Furthermore, the declinist believes that the various strategies that are usually proposed for making things better – the promotion of liberal democracy, technological development, and economic growth – cannot be the solution to our problems, since they are actually the cause. That is, it is the principles that underwrite modernity itself that are the problem. As the declinist sees it, the rights-based politics of liberal individualism, combined with the free-market economy, have served to undermine local attachments and communitarian feelings, leading us to seek meaning in the shallow consumerism and mindless entertainment that is leading us to ruin.

For a *New Yorker* article, writer Ben McGrath spent some time hanging out with members of various declinist movements in the American northeast, many of whom know one another through the Doomers.us bulletin board. He made the acquaintance of a software engineer from Russia who now lives with his wife in a Boston harbor on a sailboat that he has stuffed with beans, rice, and propane; a Connecticut commodities trader turned survivalist; some members of the Vermont independence movement; and a failed novelist turned smart-growth activist. What these people have in common is that they are all convinced that the negative aspects of the modern world – the disenchantment, the individualism, the free-market consumerism – are not just aesthetically displeasing or spiritually depressing or morally objectionable. As far as the declinists are concerned, they are going to destroy humanity.

From Oswald Spengler to Al Gore, declinism has attracted many noteworthy adherents, but perhaps the single most famous contemporary declinist is a British gentleman who made his early mark as a critic of modern architecture and who later turned is attention to land conservation and organic farming. In recent years his thinking has evolved into a full-bore critique of the modern world: modernity is a "curse" that "divides us from nature." His name is Charles, Prince of Wales, and he is first in line to the British Crown.

In a series of speeches, interviews, and articles over the past few years, Prince Charles has explained how he has come to see how his early views on architecture, the environment, and society are all tied together by a single unifying idea that he calls "the need for harmony." What undermines harmony is a mechanistic worldview that puts humanity at the center of creation, sees technology as the locomotive of progress, and fuels a disconnection that permits us to plunder the earth in the name of the "freedom it brings us, not to say the profit." As a result, he writes:

> Our perception of what we are and where we fit within the scheme of things is fractured. This is why I consider our problems today not just to be an environmental crisis, nor just a financial crisis. They all stem from this fundamental crisis in our perception. By positioning ourselves outside Nature, we have abstracted life altogether to the extent that our urbanised mentality is out of tune with the key principles underpinning the health of any economy and of all life on Earth. And those principles make up what is known as "Harmony."

What is useful about this short essay is the way Prince Charles hits every classic authenticity plot point, including the disenchantment of the world, our excessive individualism and consumerism, our obsession with technology and profit, and our inevitable alienation from nature. In order to recover from this alienation and restore our lost authentic wholeness, we need to learn "the grammar of harmony," restore our lost "balance," and achieve "organic order," by inventing technologies that "work with the grain of Nature rather than against it."

What any of this means exactly, by way of policies, institutions, or technologies, Prince Charles does not say. It is typical of this genre of critical declinism that any positive program must remain unstated, and any concessions to the benefits that have accrued to humanity over the past hundred years or so must be grudgingly downplayed or even denied. And true to form, Prince Charles does concede that while there may have been some worthwhile advances in the preceding centuries (steam trains perhaps, or maybe the Restoration), the twentieth century has been an unmitigated disaster. So the fact that life expectancy in Britain in 1900 was forty-six for men and fifty for women, compared to seventy-three and seventy-nine by 1990 – comparable to figures for the rest of Europe and North America – is of no real importance when compared with the horrors of our lack of balance with nature. When it comes to appreciating advances from indoor plumbing to central heating to antibiotics and instant communications, well, the comfort and well-being they provide pales against the alienation that ensues.

It is essential to the declinist worldview that the entire system is under indictment, with any particular stress or crisis able to stand as indicative of the underlying rot. And so in the last decade alone, declinists of various stripes have pounced on Y2K, global warming, 9/11, swine flu, and the Wall Street meltdown of 2008, holding each up in turn as evidence that the collapse is at hand. One of the most disturbing aspects of the declinist or "doomer" movement is the giddy delight with which many of them await the coming global catastrophe. So if Prince Charles does have one thing going for him, it is that unlike many declinists, he doesn't greet the impending collapse of civilization with unmitigated glee; the prospect actually seems to bother him.

The way the logic of doomer thinking works out, the declinist wins no matter what happens. We either adopt more energy-efficient, low-impact, "human scale" lifestyles or the atmosphere will heat up, the economy will collapse, and we'll be forcibly thrown back into a subsistence economy. As the great Canadian pessimist philosopher George Grant once wrote, fate leads the willing and drives the unwilling, and the declinist sees us as headed for a twelfth-century economy whether we like it or not. It's a future that the reigning high priest of declinism, James Howard Kunstler (author of *The Geography of Nowhere* and *The Long Emergency*), can hardly wait for.

Kunstler is an intellectually promiscuous declinist. He began by railing against the car culture of suburbia, but switched to Y2K in the late 1990s when it promised a more rapid apocalypse. When that didn't pan out, he switched to global warming, then to oil, but most recently it is the carnage on Wall Street in the wake of the credit crisis of 2008–09 that has given him renewed faith in the ability of humanity to destroy itself.

Kunstler likes to refer to the modern world as "a giant cluster-fuck," and when oil prices hit $150 a barrel in the summer of 2008 he wrote on his blog, "let the gloating begin." When prices quickly fell back to under $40 a barrel, he was naturally unperturbed. What matters to his vision is not that any specific calamity come to pass, whether it be the end of fish stocks or the calving off of the arctic sea ice or the freezing up of credit markets – all of these are just signs of our profoundly alienated state.

We probably shouldn't be too hard on Kunstler, since the deep relish with which he anticipates the end of the world is widely reflected in our popular culture. Michael Crichton's entire literary and cinematic output, from *The Andromeda Strain* to *Jurassic Park*, is usually classified as science fiction, but a better term for it would be declinist fiction. Director Roland Emmerich is a one-man declin-ism factory, with films such as the global warming disaster epic *The Day After Tomorrow* and the apocalyptic *2012* (which is, naturally, about the end of the Mayan long count calendar). Then there's Andrew Stanton's family-friendly declinism of *Wall•E*, which is about a lonely garbage robot stuck on an abandoned Earth. *Wall•E* touches on pretty much every declinism theme you can think of – poor eating habits, reliance on corporations, lack of consideration for the earth, technological overrun, the creeping inability to talk to people face to face – and yet not only was it nominated for an Oscar, it was also one of the most universally loved films of 2008.

One of the most enduring developments of declinism in popular culture is the ritualized destruction of the great cities of the world in film, literature, and art. Whether it is worries over economic dislocation, fears of urban alienation, or inchoate anxieties over moral and spiritual softness, we like to take it out on cities such as London, Tokyo, Washington, but, above all, New York. Historian of architecture Max Page wrote an entire book about the portrayals of New York's destruction, on paper, film, or canvas over the past hundred-odd years, showing how each era uses the city's death as a way of defining its social concerns and exorcizing its specific demons.

There's a common thread that underlies it all, though: the deadening of experience in advanced society, the banality of everyday life mixed with the precariousness of the capitalist economy. And so we use our art to destroy New York, "to escape the sense of inevitable and incomprehensible economic transformations . . . to make our world more comprehensible than it has become." Page goes on: "A disaster, even when mediated through images or words, still retains an authenticity that has been the quest of modern society for two centuries."

But why New York? A clue is to be found in the way in which, in the years after 9/11, the attack on the Pentagon has almost completely faded from popular remembrance. Washington, D.C., may be the capital of the American empire, but New York is the capital of modernity, or as Oswald Spengler put it, the "monstrous symbol" of the modern world. Whether it is King Kong making his final stand atop the Empire State Building or the lizard in *Cloverfield* ripping the head off of the Statue of Liberty, it is something significantly more than a tourist attraction that is under assault from these monsters of nature.

The primitivist line of thinking that stems from Rousseau's unbending critique of modernity, leading through the nineteenth-century obsession with the exotic and up to the gleeful declinism of today, reflects an enduring desire to deal with the causes of our alienation by reverting to a simpler form of social and economic organization. Recall though that when it comes to dealing with the negative effects of modernity, we have a choice: we can either try to eliminate the causes of our problems or we can try to mitigate the effects.

The irony is that Rousseau himself took the second tack, preferring to cope with the modern world on its own terms. He looked to dampen the poisonous effects of society by building stronger, more resourceful and autonomous individuals. He didn't want to see our cities destroyed, and he had no real desire to see humans retreat into some quasi-tribal state of nature. No, Rousseau didn't want to go into the woods; where he really wanted to go was back to his hometown of Geneva, where keeping it real amounted to little more than hanging with friends, doing some hunting, and playing a bit of backgammon. Rousseau was no primitivist, far from it. More plausibly, he was the prototypical bohemian whose true descendants include William Wordsworth, the American transcendentalist writers such as Henry David Thoreau and Ralph Waldo Emerson, and eventually the Beat poets, the hippies, and the flourishing counterculture of the late twentieth century.

Think again of Rousseau's little story about the comb, which, he claims, marked the moment he lost his innocence and discovered the horrors that can arise from being falsely accused. Doesn't it sound just a little too pat, or even contrived? Even assuming such an incident ever took place, doesn't Rousseau's subsequent interpretation seem like a heavily torqued bit of biographical revisionism? Perhaps. But as it turns out, for the account of individual authenticity that Rousseau is trying to establish, the historical truth of what happened is not really relevant. Far more important is how Rousseau *feels* about the incident and how it continues to color his self-understanding.

To recap: the central concern of Rousseau's philosophical project is to distinguish what is natural from what is artificial in the state of men in society. He knows that civilization deforms human nature, but the precise contours of that deformation are unclear. Another way of putting it is to say that what Rousseau is trying to do is answer the question Who am I? And who he really and truly is can only be determined once he is able to strip away the masks and role-playing of social life, pull back from the petty competitions and game-playing, and search for what he truly wants outside the false demands of society.

And so the popular, primitivist view of Rousseau's ambition is mistaken: instead of looking for some sort of modernity-free sanctuary out somewhere in the world or in our distant past, he proposed that we look inward and find our authentic self by attending to our most basic, spontaneous, and powerful feelings and emotions. In this view, the authentic person is someone who is in touch with their deepest feelings, whose emotional life is laid bare. It is by listening to what my feelings have to say that I can discern what is most truly me. Who am I? Rousseau knew the answer to that: *Je sens mon coeur*, he writes, "I feel my heart." Or as the philosopher Charles Guignon puts it, Rousseau's answer to the question is, "I truly *am* what I feel myself to be."

This way of interpreting what Rousseau was up to makes him seem absurdly self-centered. But whatever else it may be, an authentic self has an irreducibly social dimension. Go back to the story of the comb: it is not enough to be transparent and honest with oneself – one has to be recognized for that honesty by others. That is why, in addition to his works of social criticism and his lengthy treatise on education, Rousseau also produced a huge amount of autobiography, in which he indulged his insistent desire to tell *other people* the unvarnished truth about his life.

There is a problem though. On the one hand, Rousseau wants to tell the raw and unembellished facts of his life, to produce a clear representation of who he is and what he is about. But he also knows that we don't have unmediated access to the past; it is always filtered through our present emotional state, which shapes and colors our interpretation of what has come before. The truth is an elusive beast, and one that ultimately Rousseau does not think is worth pursuing. After all, why chase after slippery facts when our feelings are always ready at hand? As he writes in his *Confessions*: "I have only one faithful guide on which I can count: the succession of feelings that have marked the development of my being. . . . I may omit or transpose facts, but *I cannot go wrong about what I have felt* or about what my feelings have led me to do."

What Rousseau is doing here is incredibly bold. He takes the Cartesian search for certainty and completely upends it, so where Descartes concluded that the search for truth could only begin with an indubitable fact ("I am, I exist"), Rousseau says, forget the facts, they'll only lead you astray. Truth begins with the indubitability of emotions, and only once you know how you feel can you make any progress. This is a seismic shift in philosophical foundations, and its reverberations are still being felt throughout our culture. The idea that one's feelings have this privileged position with respect to the truth marks a fateful turn in the search for the authentic self, away from the discovery of objective facts toward a subjective focus on our personal emotions and commitments. Along the way, the answer to the question Who am I? is itself transformed from a process of discovery to something closer to an act of creation. We don't find our authentic self by peeling away the shell of civilization until we reach the hard nut of the natural self at the core. The self is more like an onion; there is no "natural self" to be found at the center because there is no center. Authenticity becomes redefined as the ongoing process of filtering our experiences through our most deeply felt emotions and constantly interpreting and reinterpreting our lives until we find a story that is uniquely our own. So in the end, who cares whether Rousseau is lying about the comb? What matters to the story is the role it plays in Rousseau's current self-understanding, as the catalyst for his entire quest to be seen as someone who hides nothing, puts it all on the table.

The reason this is such a fateful turn is that it firmly establishes the quest for the authentic as an artistic enterprise. Being true to yourself, in the sense that Polonius intended it, is now a lifelong creative project from which no one is exempt, and it plants the solitary artist at the center of our moral understanding. This is the Romantic turn in the modern worldview, heralding the start of a backlash against science, rationalism, and commerce. The authentic individual is one who disengages from the deforming forces of society and looks inward, drawing inspiration from the murky depths of the creative self. The turn inward is a quasi-religious quest, where the creative powers of the individual – *not* its power of reason – come to be seen as the last true source of meaning in a world that is otherwise sterile and disenchanted.

We can safely say that it was Rousseau who launched the first serious volley in the culture wars, the now centuries-long dispute between passion and reason, art and commerce, the individual and society, the bohemian and the bourgeois. To be bourgeois is to be alienated from your authentic self, which is just another way of saying that you've allowed your creativity to atrophy in the name of comfort and security. You've sold out, in other words, and the only way to get your edge back is to become a bohemian, a nonconformist, a solitary rebel at odds and out of step with the mainstream. An authentic person is one who, almost by definition, rejects popular tastes, thoughts, opinions, styles, and morals.

Rousseau laid the groundwork for our understanding of the authentic self and its relationship to the modern world, and most of what is to come in this book will involve exploring the consequences of these developments and showing how the basic terms of engagement that he laid down are still the ones that dominate our approach to the questions of personal identity and the meaningful life.

—

THREE

THE CREATIVE SELF

Joseph Wagenbach was a German who immigrated to Toronto in 1967. He spent the next forty years as a recluse, keeping mostly to himself in his small house in the city's west end. In early 2006, he suffered a massive stroke and was hospitalized. Unable to contact any friends or relatives, city officials took control of his affairs and prepared to put the house up for sale. But when they went inside, they found a massive cache of his art that included paintings, sketches, sculpted figurines, and found objects. An archivist for the city was captivated by what they had discovered and set about trying to piece together Wagenbach's life.

The archivist, Iris Häussler, was determined to have the house and its contents designated a historical site. So she and her assistant began taking people on walking tours through the house, giving lectures about Wagenbach's life and art, trying to drum up public support for the project. Visitors passed his hat and tweed jacket hanging in the entrance, then went through a dirty kitchen to an improvised gallery at the back, which displayed sculpted nudes, teddy bears, and columns of stacked flower pots glued together.

The most heartbreaking room in the house belonged to Wagenbach's former lover and muse, Anna Neritti. He had the room sealed in 1974 after Neritti left him, and it remained untouched for more than thirty years. In it was a pre–Second World War map of Germany marked with the locations of concentration camps, leading many to suppose that his reclusiveness was in some way connected to the Holocaust. All in all, it was a tremendously moving exhibit, raising many important questions about the nature of art and how it connects to biography, loneliness, and love.

And an exhibit is exactly what it was. You see, Joseph Wagenbach did not exist. His life, his art, indeed the entire contents of the house off Queen St. West in Toronto, was the creation of Iris Häussler, artist. Everything, including the City of Toronto Archives sign on the front lawn, was part of an installation put together by Häussler. That is, the Wagenbach house was an elaborate hoax.

The various ways in which we react to artistic forgeries and frauds helps tease out our assumptions about art: about what counts as art and why we care about the origins of a work. After touring the Wagenbach house, one journalist wondered whether the whole point of the installation was to provoke a raw and unmediated aesthetic reaction in the viewer. But if that was Häussler's agenda, then it isn't clear why she would have concocted the whole backstory, instead of simply presenting the house to the public without any explanation at all and challenging people to react to it.

Instead, what Häussler was doing was playing with the past in order to show how our response to a work of art is conditioned by our assumptions about the artist's biography. Instead of providing a privileged look into the private world of a lonely old man haunted by the past, Häussler forces us to deal with the all-too-familiar modern reality of manipulation and deceit, along with the uneasy realization that the artist has a hidden agenda. What it comes down to is the question of motive: "I want to believe there are things going on in Toronto that defy explanation and have escaped commodification," the journalist wrote. "I like to think that a man around the corner from my house locked himself away to create work only for himself, fed by the spirit of a woman he had loved and lost." What she is really wondering is, Is this work of art authentic?

In the last couple of chapters, we have followed the turn in Western culture that began with an initial, visceral reaction against the three pillars of the modern world: spiritual disenchantment, political liberalism, and the growth of the market economy. As we traced it through the thought of Jean-Jacques Rousseau, this reaction gave rise to the ideal of authenticity, which culminated in a celebration of spontaneity, emotional transparency, and a fixation on the creative powers of the individual to provide meaning in a world that otherwise offers none.

This last development is particularly important. Once the authentic self becomes, in effect, an artistic project, that puts a number of questions relating to art and authenticity front and center. What counts as an authentic work of art? What threatens artistic authenticity? How can we tell the difference between art that is genuinely authentic and works that only seem so? Or is that last distinction even meaningful?

There is an ambiguity in the way we use the term *authenticity* when discussing art. The first kind of authenticity, what the art world refers to as its provenance, is concerned with the correct identification of the origins or authorship of an object or work.

Sometimes, though, the provenance of an object is not in question. We know where it came from, yet we are still beset by concerns about its authenticity. On these occasions, we are worried about a discrepancy between what the artist seems to be doing and what he or she is actually up to. That is, we look at a work and we wonder whether the work is a true expression of the artist's *self*, her *vision*, her *ideals*, or perhaps her community, culture, or "scene." What we are concerned with in this case is that there is a divergence between the art that is expressed and what we think the artist ought to be expressing, or is entitled to express.

The connection between authenticity-as-provenance and authenticity-as-artistic-expression is not immediately apparent. The first is a matter of empirical facts about the history of a work, and it should not present us with any great philosophical difficulties. However hard in practice it might be to determine, surely it is simply a matter of fact that either the St. James Ossuary once contained the bones of Christ or it did not. Expressive authenticity, on the other hand, is concerned not with facts but with values. When we ask whether a work is authentic in this sense, we are being asked to judge the degree of fit between an artist's true self and the work he or she has created, or between an artistic work and the community or culture it purports to represent. Was *Never Mind the Bollocks: Here's the Sex Pistols* a true (that is, authentic) expression of the London punk scene of the late 1970s, or was it a work of cynical opportunism by Malcolm McLaren?

When we scratch the surface a bit, it turns out that these two types of authenticity aren't as unrelated as they initially appear. In fact, it turns out that our determinations about the expressive authenticity of a work of art – be it a painting, a song, or a novel – often depend quite heavily on its provenance. Whether we judge something authentic in the expressive sense turns out to depend on the artist's background and what his or her intention was in making the work in question.

In early 2002, a teenage singer named Avril Lavigne came out of nowhere (well, Napanee, Ontario, so close enough) to dominate the pop charts. The first single off her album *Let Go* was called "Complicated," and it reached number one in a bunch of countries, peaking at number two in the United States. Two subsequent singles, "Sk8r Boi" and "I'm With You," also sold well, both making it into the top ten stateside.

Almost immediately, Lavigne was subject to the usual suspicions about who actually wrote her songs. She claimed that they were all her compositions, but the confidence of the songwriting and the maturity of the lyrics was belied by the painful naïveté, bordering on inarticulacy, that the seventeen-year-old Lavigne displayed in interviews. But what most people cared about was not whether she had written the songs on her record, since they were quite obviously written by her production team, The Matrix, the multiple-Grammy-winning trio that has written songs for Christina Aguilera, Ashley Tisdale, Shakira, Liz Phair, Korn, and others. Instead, a vile rumor swept through the chatrooms and listservs of the Internet: *Avril Lavigne doesn't know how to skateboard.*

This was a far more damaging accusation. Lavigne's whole artistic persona (at the time) was built around the proposition that she was a real-deal skater chick from the streets of small-town Ontario. She certainly looked the part, with her baggy army shorts, skater shoes, flannel shirts waving over torn T-shirts, and she made a point of being photographed carrying a skateboard around. And so while she maybe didn't actually write the song "Sk8r Boi," there was no question that the lyrics – about an alienated young skateboarder/musician who gets famous and hooks up with a girl who really understands him – were supposed to express some fundamental truths about the world Lavigne came from. If it turns out that she can't even point the deck in the right direction, well, that changes everything, doesn't it? It's like finding out that the Beach Boys didn't know how to surf.

The underlying intuition here is that there is an intimate connection between your upbringing and your identity; that the biographical question "Where are you from?" is a reliable guide to answering the existential question "Where are you coming from?" Further, there's a normative dimension to this, insofar as your background (including your race, your class, your schooling, even what part of the country you are from) frames the scope and limits of what you can legitimately claim to speak, or sing, or paint, or write about.

This intuition manifests itself all over the place. For example, it is what drives one of the longest-running battles in the culture wars, over "appropriation of voice" and the question of when, if ever, it is permissible for someone of one culture or racial background to speak in the voice of another. It is usually flagged as a matter of concern when a member of a dominant group adopts the purported experiences of a dominated group as a way of making a political statement, such as when white cultural studies students write from the perspective of poverty-stricken Aboriginals or when men try to write about female oppression.

Such appropriations are usually deemed racist, sexist, classist, or what have you, but as U.S. Supreme Court Justice Sonia Sotomayor found out, the debate does cut both ways. After President Barack Obama nominated Sotomayor, of Puerto Rican descent, to the court in 2009, it emerged that a few years previously she had given a speech in which she had said she hoped "a wise Latina woman" would be able to decide cases better than a white male judge who lacked the "richness of her experiences." Sotomayor was quickly denounced by conservative pundits as a "reverse racist" (to which Rush Limbaugh added that she was a "hack"), and she was widely condemned as unfit for the position.

The racism charge is pretty dubious, and Sotomayor sounded sincere when she confessed during her confirmation hearing that the phrase was nothing more than a failed rhetorical device aimed at encouraging young Latin Americans to go into law. More than anything, what the Sotomayor incident highlights is the way this type of identity politics quickly turns into a form of status competition, where the relative authenticity of one voice over another results in a game of moral one-upmanship.

A more entertaining version of this lesson is found in *8 Mile*, a quasi-biographical film about the white rapper Eminem. It is basically *Rocky* transposed into the realm of rap music, and instead of boxing matches, the plot is driven by the desire of the main character, B-Rabbit (played by Eminem), to prove himself in a series of head-to-head rhyming contests against other rappers. Rabbit's Achilles' heel is the fact that he's a white kid facing off against mostly black kids from the mean streets, on their turf and in their language. But in the final showdown, Rabbit does a bit of hip-hop jiu-jitsu on his opponent by confessing all his biographical sins, including the fact that he's white, lives in a trailer, and his girlfriend cheated on him. But then Rabbit lets the crowd in on a secret: his opponent might be black and good at striking a gangster pose, but – get this – his parents are still together. Worse, he went to a private school, and his real name is Clarence. Rabbit wins the bout, his hip-hop authenticity trumping that of "Clarence" in a knockout.

There's nothing terribly new about this dynamic. Take popular music, where virtually every artist of note over the past fifty years has had their expressive authenticity called into question. Elvis was just a white boy stealing from the real blues musicians, and the Monkees were just aping the Beatles. The Sex Pistols didn't emerge organically out of the London punk scene of the 1970s but were created out of whole cloth by fashion impresario Malcolm McLaren. And so on, down to bands such as the Jon Spencer Blues Explosion and The Strokes, who were both dismissed by some critics as consisting of trust-fund babies who hadn't earned the gritty street cred advertised by their music.

The name for this is co-optation. According to the standard picture of cultural co-optation, what happens is an authentic art form emerges organically out of a given subcultural milieu. Eventually, members of the dominant culture (usually rich white males) come along and appropriate the superficial looks or sounds or techniques of this artform while taking some sandpaper to its rougher edges. This softened version is then sold to the masses as the real thing, while the true innovators of the art form are left to starve in obscurity.

But set aside the moral question of whether artistic innovators should receive due compensation from those who appropriate their work. That's certainly important, but there's a deeper question here about the metaphysics of aesthetic judgment: What happens if we can't tell the difference between the original and the fake, or between the authentic and the ersatz? After all, an artist's biography is not self-revealing, and it is not always possible to tell an object's history just by looking at it. But if we can't tell the difference just by looking or listening, what difference could it possibly make?

When it comes to art, religious objects, and cultural artifacts, fakes, forgeries, and hoaxes are probably as old as humanity. In his book *False Impressions: The Hunt for Big-Time Art Fakes*, former director of the Metropolitan Museum of Art Thomas Hoving traces the practice as far back as the Phoenicians, who routinely trafficked in art fakes, especially phony Egyptian pottery. Yet the Phoenicians had nothing on the ancient Romans, who produced countless forgeries of Greek artifacts, who in turn couldn't hold a candle to the innumerable phony crucifixion relics from the Middle Ages and the phony sculpture, paintings, documents, coins, ivories, and gems created largely for profit. In more ways than one, "faking it" may indeed be the oldest profession.

As Hoving's thorough account makes clear, people concoct fakes for a bunch of different reasons. Money is the most obvious motivation, with phony sculptures, paintings, coins, and figures created largely for profit. But forgeries are also created to promote a political or religious agenda (as in the case of the Shroud of Turin) or to satisfy a personal vendetta (the famous van Meegeren Vermeers). It isn't even true that primary goal of the forger is to deceive. Many artists make copies of famous works as a way of practicing a certain technique, or as an exercise in creative mimicry. Still others are providing a service similar to the countless cover bands that mimic famous musical groups, right down to the costume design and stage show: when the real version is either too expensive or not available at any price, sometimes people are willing to settle for a reasonable facsimile.

Because fakery differs in its intentions, it is useful to draw some distinctions. Let us define a *forgery* as a work of art whose origins or history are deliberately misrepresented by someone to an audience, normally for financial gain. Similarly, a *hoax* is a work of art that is misrepresented as having a different history or origin than it actually does. What distinguishes a forgery from a hoax is that, in the case of forgery, the work is falsely attributed to a known artist. It is rare to find someone trying to sell an exact copy of a known work (for obvious reasons, it would be a tough sell), which is why forgeries are usually created "in the style of" an artist whose complete body of work is relatively unknown.

What forgeries and hoaxes have in common is that in both cases, the misrepresentation is deliberate, unlike in cases of mistaken attribution. These are instances where a dealer or a curator simply gets it wrong, misidentifying a work and thereby claiming that it has a past different from the one it actually has. Of course, there is a fine line between innocent misattribution and deliberate misrepresentation, and some find it convenient to fudge the truth, offering overly optimistic estimations of the age or quality of a work.

The obvious financial considerations are always at work here, but there is a moral dimension as well. In a famous remark attributed to legendary German art historian Max J. Friedländer, "It is indeed an error to collect a forgery but it is a sin to stamp a genuine piece with the seal of falsehood!" This is the curatorial equivalent of the legal maxim of "innocent until proven guilty," backed by the conviction that it is better that a hundred criminals go free than an innocent man hang. Still, all things considered, we would rather not have our galleries crowded with undetected forgeries any more than we would delight in having our streets overrun with unconvicted criminals. Almost always, forgeries and fakes are poorly done and quickly discovered, revealed as cheap publicity stunts or clumsy attempts at bamboozling the guileless or the ignorant.

More often than should make us comfortable though, even the experts are fooled. One of the most famous examples is the case of Han van Meegeren, a forger of Vermeer. Van Meegeren was a failed artist, dismissed in his time, who took to faking Vermeers as a way of exacting revenge on the art world that had rejected him. He had the pleasure of seeing dozens of his paintings accepted as genuine Vermeers, culminating in the sale of a painting called *Emmaus*. At the official unveiling of the painting, van Meegeren stood in the audience as the German curator declared the work to be not just an authentic Vermeer, but a masterpiece. Van Meegeren was only caught when he was prosecuted by the Dutch authorities for illegally selling national treasures to the Germans. He was forced to confess that he had forged the Vermeers, proving it by painting in his cell while awaiting trial.

While they aren't all so colorful, the history of what Thomas Hoving calls fakebusting is full examples of so-called experts or authorities being completely fooled. This certainly calls into question the validity of the expertise, but it also raises more serious questions about the aesthetic values that are being judged in the first place. Didn't van Meegeren make his point, that his art *did* deserve to hang in galleries beside the work of the great Old Masters? After all, if even the experts cannot tell the difference between a van Meegeren and a Vermeer, why should the rest of us care?

At the limit, the convincing forgery leads to the problem of the perfectly convincing copy. If we have two works, one original, the other a copy so competent that no one can tell the difference between the two, should it matter to us which one hangs in the gallery? Almost everyone believes that it matters a great deal. Not only do tens of millions of dollars hinge on these sorts of questions, but so does an entire realm of aesthetic experience. It is central to our concept of authenticity that the past infuses the present, and that the mark of the authentic work of art is that it has, by some measure, the right origins and the right history.

In establishing the provenance of a work, art experts rely on four main tests: verifying the signature of the artist on the work; reviewing the historical documentation, or paper trail, that guarantees the work's history; studying scientific evidence that can include X-rays of the canvas, infrared spectroscopy of the paint, and dendrochronological analysis of the panels; and an expert judgment by a trained eye, known as the *connoisseur*.

In the popular imagination, there is a tendency to assume that proper authentication relies on the first three tests – the signature, the provenance, and the laboratory analysis. To a certain extent, this no doubt reflects a widespread skepticism about "expert opinion" and the bizarrely unsubstantiated claims of the connoisseur. And the public's faith in the supposed objectivity of scientific evidence plays out in other realms, such as the belief that the best evidence for evolution by natural selection is found in the fossil record or that the best way to convict an accused criminal is with fingerprints or DNA evidence. The great cultural expression of this faith is found in what appears to be an unlimited appetite for television devoted to forensic medicine. By 2009, North American viewers could choose from no fewer than fifteen fiction and nonfiction shows about forensics, with barely distinguishable names such as *Cold Case*, *Cold Squad*, *Cold Case Files*, *Extreme Evidence*, and of course the CSI franchise, with shows set in Las Vegas, Miami, and New York. What all these shows share is an obsession with the idea that the telltale clue will be revealed only under the gaze of a scientific instrument. Similarly, it is tempting to believe that art fakes will be revealed through details that the forger has missed, tracks he has forgotten to cover.

Yet all three of these tests have serious defects when it comes to helping determine the authenticity of a work. Besides the fact that signatures are relatively easy to fake, many artists hide their signatures in the work, or use them inconsistently or not at all. With respect to the provenance, there is no question that it can occasionally provide the equivalent of a smoking gun, such as a photo of the artist actually creating the work. But for a great many works, the provenance is marred by gaps in the ownership record and by obscure, incomplete, and ambiguous documentation. Worse, there is a large market for forged certificates of authenticity, provided for a fee by unscrupulous dealers. The problem is so bad that a rule of thumb in the art world is that the more complete and airtight the provenance, the more suspicious the work.

Even scientific evidence is less useful than we might imagine. Despite the faith that many people have in science, laboratory results can be ambiguous or even contradictory. Sure, there is the occasional glorious success, such as the time an X-ray revealed a sketch of the Eiffel Tower underneath a painting that was purportedly a seventeenth-century Rubens. But even at its most useful, technology is only able to provide negative evidence, by exposing fakes. That is, it can disprove a claim of authenticity, tell us that a work is most definitely not a Rembrandt, for example. But it cannot tell us what it actually is. For that, we need the expert eye of the connoisseur.

There's something a bit mysterious and unscientific about the whole notion of connoisseurship, and the air of mumbo-jumbo that surrounds it is not dispelled by the explanations of the connoisseurs themselves. Thomas Hoving says that the best connoisseurs have a sixth or even seventh sense that lets them "instantaneously" detect a forgery in any field, but what these additional senses consist in he does not say. Here is his account of the talents of Bernard Berenson (1865–1959), the great Lithuanian-American historian of Renaissance art:

> He sometimes distressed his colleagues with his inability to articulate how he could see so clearly the tiny defects and inconsistencies

in a particular work that branded it either an unintelligent rework-
ing or a fake. In one court case, in fact, Berenson was able to say
only that his stomach felt wrong. He had a curious ringing in his
ears. He was struck by a momentary depression. Or he felt woozy
and off balance. Hardly scientific descriptions of how he knew he
was in the presence of something cooked up or faked. But that's as
far as he was able to go.

Yet for all the impressionistic hand-waving at what it is about, connoisseurship is rooted in techniques of inquiry that are almost as
objective and scientific as the X-ray machine.

The most important element of connoisseurship in the visual
arts is the notion of form. For the connoisseur, form is not just
about identifying the gross shape or structure of a work, but is
instead something much more personal. Form, according to one
scholar, "is the manner or personal style of the artist that deter-
mines the characteristics and strategies which comprise his or her
visual temperament." Form is the equivalent to the novelist's
"voice" or the musician's "sound." It is the distinctive use of vocabulary, phrasing, tone, and so on that makes a work uniquely theirs.

(It is important to note that form is not the same as style.
Cubism and Impressionism are styles, but each Cubist or
Impressionist had their own individual form. Furthermore, an
artist can change styles while their underlying form remains more
or less constant – think of Picasso moving from his Blue to his
Rose Period, or Sting moving from punk to pop to lite jazz. Nor is
form a function of quality, since an artist's form shines through
even his most poorly executed works.)

In trying to identify a work by recognizing the artist's distinctive
form, the connoisseur becomes something like a private investi-
gator who realizes that the best way to catch a criminal is to learn
to think like one. The connoisseur must be so familiar with the
way form is expressed in a work that he is able to psychologically

identify with the artist. The expert connoisseur must know how the artist thinks and feels, must understand his artistic vision and temperament. He must, in some sense, identify with the artist. What the connoisseur does, in effect, is treat the entire work of art as a signature. Using form as an authentication technique is far more reliable than looking at signature and scientific methods. After all, what is more difficult, copying Jackson Pollock's signature or teaching yourself to paint like Jackson Pollock?

The notion of the connoisseur as the Sherlock Holmes of the art world was firmly established in the latter half of the nineteenth century, through the work of Italian physician and art critic Giovanni Morelli. As far as Morelli was concerned, the art criticism of his time was virtually useless, and he was determined to distinguish his method of connoisseurship from the prevailing canons of art history. He felt that his contemporaries gave too much weight to irrelevant types of evidence, such as historical attributions, documentation, and aesthetic judgment, Morelli argued that true connoisseurship must attend to the precise details of the painting itself.

Morelli believed that the telltale signs of authorship could be discerned not in the historical record, nor in any high-minded appeals to the work's "aesthetic essence," but in the insignificant details such as the curve of a fingernail or an earlobe, or the fall of a drapery. It is in the quick and inattentive renderings of stereotyped elements that the artist reveals his identity, because it is when we are at our most natural and unconscious that we are most ourselves: "Just as most men, both speakers and writers, make use of habitual modes of expression, favorite words or sayings that they employ involuntarily, even inappropriately, so too every painter has his own particularities that escape him without his being aware of them."

Morelli laid out his method in two groundbreaking texts, *Italian Masters in German Galleries* (1880) and *Critical Studies of Italian*

Painters (1890). At the heart of it was the laborious construction of charts of stereotyped forms for dozens of artists, page after page of ears and hands by Filippino, Botticelli, Giorgione, and so on. Using these charts, he was able to compare repeated motifs in known works by an artist with similar motifs from a new painting. It was essentially a form of fingerprint analysis applied to painting, and it paid off with some spectacular discoveries. He didn't always get it right though, and the Morellian technique was soon criticized for being too rigid and formulaic.

For the most part, these criticisms only took issue with the sophistication of the technique's "grain." They did not challenge the underlying assumption: that an artist reveals his identity through his form, which can be determined in a rigorous and objective manner. Thus, the crucial insight (or rather, supposition) of contemporary connoisseurship is that there is a fusion of the two kinds of authenticity – the provenance and the expressiveness. The idea is that a work of art is in some sense pregnant with its past and the artist's true creative spirit shines through. How can you tell Patsy Cline had a tumultuous personal life? Just listen to the emotion in her singing. Isn't it obvious that Mark Rothko was suicidal? Stand in front of the Seagram murals for a spell and you'll feel it too.

Except it isn't always so simple. We might think it is a straightforward, empirical fact whether a painting is an authentic Rembrandt, and the connoisseur is the one who can tell us. But in a world where art can be copied, reworked, and reproduced in an indefinite number of copies, the very idea of the "original" work becomes problematic, and by the end of the twentieth century it had led to a serious crisis of authenticity in the world of art.

London, 1990. The advertising magnate and art patron Charles Saatchi is standing, incredulous, in front of a large glass case, inside which is the rotting head of a cow being slowly devoured by maggots.

Saatchi promptly buys the work, entitled *A Thousand Years*, and then makes a blanket proposal to the artist. Saatchi is willing to foot the bill for anything Damien Hirst wants to make next.

What Hirst wanted to make next involved putting a fourteen-foot tiger shark in a large tank, pickling it in formaldehyde, and calling the whole thing *The Physical Impossibility of Death in the Mind of Someone Living*. It cost Saatchi £50,000. Next, Hirst presented a cow and her calf, cut into pieces, and called it *Mother and Child Divided*, followed by *Away from the Flock*, which was a dead sheep in a tank.

The shark in the tank drew a predictable (and for the artist, desirable) level of outrage from art critics and the popular press alike. "No more interesting than a stuffed pike above a pub door," said one critic. "50 000 for fish without chips" screamed *The Sun*. Of course, the art world being what it is, the outrage soon turned to lavish praise, and Hirst won the Turner Prize in 1995 for *Mother and Child Divided*. That same year, the piece *Two Fucking and Two Watching*, which featured a rotting cow and bull, was banned by New York health authorities because of fears of "vomiting among the visitors." Hirst has gone on to become one of the wealthiest men in Britain.

But there was a problem with the shark. Because it was not properly preserved, it soon changed color, got all wrinkled, and began to decompose, with the water in the tank going all murky. A year after he bought it, Saatchi's curators decided to skin the shark and fix the skin over a fiberglass mold, though Hirst never liked the effect. When Saatchi sold the piece in 2004 to the American hedge fund billionaire Steven Cohen, for US$8 million, Hirst offered to replace the shark. This time, he had the shark professionally preserved, with formaldehyde being injected into every cell in the new shark's body.

While *The Physical Impossibility of Death in the Mind of Someone Living* is no longer a scandal, the reconstituted version will cause a new problem for art critics: is it the same piece? That is, is the version of TPIODITMOSL that will take up residence in Steven Cohen's home in Connecticut the same work of art Charles Saatchi purchased in 1992? And to confound things even further, what if – unbeknownst to Cohen – Hirst took the original shark, the one now stretched over a fiberglass mold, and put it in a separate tank. Which one would we call the "real," "original," or "authentic" work? Does the question even make sense?

This is a variation on a conundrum that has annoyed philosophers for centuries, and it is known as the Ship of Theseus problem, thanks to a story told by Plutarch in his *Life of Theseus*:

> The ship wherein Theseus and the youth of Athens returned had thirty oars, and was preserved by the Athenians down even to the time of Demetrius Phalereus, for they took away the old planks as they decayed, putting in new and stronger timber in their place, insomuch that this ship became a standing example among the philosophers, for the logical question of things that grow; one side holding that the ship remained the same, and the other contending that it was not the same.

Is the ship that was preserved and renovated for years by the Athenians the same ship that brought Theseus back from Crete? Making things worse, a number of philosophers added the following embellishment to the debate. What if, unbeknownst to the Athenians, while they were gradually putting new planks in the ship, someone was collecting the old ones and using them to build a new ship? Which, if either, would we say is the real ship of Theseus?

The paradox works by tugging our intuitions in opposing directions. On the one hand, it seems that a thing is the sum of all of its properties and characteristics. If any of these properties change, then the object is, by definition, no longer the same thing, though it may be related in certain important ways to the prior object. This intuition is what underlies Heraclitus's famous argument that you can't step in the same river twice, on the grounds that it is always flowing. For Heraclitus, everything flows, nothing is stable.

Yet this is at odds with how we tend to speak about things. As a rule, it seems that the following is true: in order for something to change, it has to remain the same. After all, how could we say that something had changed unless we assumed that it had somehow persevered and continued to exist as the same thing? Socrates had hair, then he went bald, but he remained Socrates. If I paint my fence, I don't say that I got a new fence but that my old fence now has a new coat of paint.

The problem of the persistence of the individual through change is one of the fundamental problems in metaphysics, and, as is so often the case, Aristotle provides the best tools for getting a handle on things. He begins by noting that questions like "Is this the same ship as it was before we replaced all the planks?" or "Is this the same work of art before we replaced the shark?" are almost never intended in the strict Heraclitian sense. Rather, we usually qualify the question by asking, in what respects, or for what purposes, has the object remained the same?

What he is getting at is the idea that a judgment of identity is always just that: a judgment call, made in light of some analytical or practical intent. With respect to art, often we are not really worried about whether something is made out of the exact same materials, or even whether it has completely retained its previous shape or composition. Instead, what we want to know is whether it has the same expressive power that the artist intended and the same ability to evoke the desired aesthetic, intellectual, or even just emotional reaction in the audience. Indeed, this is pretty much the answer Hirst himself gave in an interview with *The New York Times*, when he acknowledged the problem: "Artists and conservators have different opinions about what's important: the original artwork or the original intention. I come from a Conceptual art background, so I think it should be the intention. It's the same piece. But the jury will be out for a long time."

So just how much change can a work undergo before its expressive power becomes compromised? The trouble stems from the basic assumption of connoisseurship, namely, that the individuality of a work can be discerned from its form. But form can change. Works can be renovated and restored, they can be altered, painted over, and "improved" by overzealous or overambitious owners. Even the accretions of time can affect the form of a work, smoothing deliberately rough edges or dulling the original surface.

As much as possible, conservators try to preserve the original integrity of a work, but there are no rules about how much change is allowed, and of what sort. In some cases, historians concede that eventually the authenticity of a work becomes a matter of degrees, leaving us with a work that is half Rembrandt, or one-third Pollock. Further complications arise when we are forced to interpret the intentions of the restorer, as in the case of a painting by Egon Schiele that was bought through Christie's in 1987 for £500,000. It was later discovered that more than 90 per cent of the work had been "overpainted" by someone who had followed the original design and used the original color scheme.

The case eventually went to court, where the judge ruled that overpainting a work, even a significant portion of it, is fine as long as the intention is to restore the original work. Yet when it was revealed that the conservator had traced over Schiele's original mauve initials with black paint, the judge felt he had gone too far. This was seen as a deliberate attempt to deceive (by making it look like Schiele had signed the "restored" version) and Christie's was ordered to refund the purchaser's money.

A more interesting difficulty arises when the artist's form is the presentation of a distinctive aesthetic effect, such as Mark Rothko's resonant fields of color. Many of Rothko's works, such as the

Harvard Murals and his fourteen "blood red on red" paintings in the Houston Chapel, have changed color over time thanks to photochemical reactions in the pigments. In the case of the Harvard murals, the paintings have had to be taken down and put in storage, their authenticity seriously in question.

Yet even these hard cases don't compare to the problems that arise when there is no "original" work at all, and when the expressive intentions of the creator are not just difficult, but downright impossible to discern.

Walter Benjamin is one of the more quietly interesting figures in twentieth-century ideas. He was a Jew born in Berlin in 1892, and his literary career spanned little more than the decade leading up to the Second World War. He is most widely known for his affiliation with the group of neo-Marxist philosophers and critics known as the Frankfurt School, which included three of the biggest heavyweights of cultural theory: Max Horkheimer, Theodor Adorno, and Herbert Marcuse.

When the Nazis came to power in 1933, Benjamin left for Paris. But as the Wehrmacht made its way toward the French capital in 1940, he fled once again, aiming to make his way first to Spain, then to Portugal, and from there to America. In August 1940, he obtained an entry visa to the United States, but for reasons that remain lost to history, he never made it. The best reconstruction of events suggests that he reached the town of Portbou, on the Spanish border in the Pyrenees. There, he appears to have committed suicide, probably taking an overdose of morphine.

Walter Benjamin left behind a large and eclectic body of work, the most important of which may turn out to be his massive (and unfinished) study of the Paris arcades of the nineteenth century. The Arcades Project was not published until 1999, and scholars have only recently started giving it the attention it deserves. Yet there is one Walter Benjamin work that every student of philosophy, literature, or cultural studies knows inside and out, and that is his essay "The Work of Art in the Age of Mechanical Reproduction." Published in 1936, the essay remains the best statement of our intuitions about the meaning of art, while helping expose our anxieties about authenticity that began with photography and film and continue today in the rip/mix/burn culture of digital collage.

Benjamin argues that there is a straightforward answer to the question of what distinguishes an original work of art from the perfect copy, since even the perfect copy is lacking in one crucial element, namely, its "presence in time and space, its unique existence at the place where it happens to be." Only the original work has that unique history, has traced that particular wormhole through space-time. Two seemingly identical objects differ at least in the respect that they have different, unique pasts.

This sounds important and profound. But hang on a second: there's nothing in the world so common as "uniqueness," since everything that exists just is what it is, occupying its own particular place in the space-time continuum. By this measure, the most valuable and irreplaceable works in the world, such as the Mona Lisa, are no more "unique" than the cheesiest poker-dogs-on-velvet print. That is why for Benjamin, the sense of awe or veneration we have for an authentic relic or a work of art is captured by more than just its past.

What we value is its *aura*, which consists in the history and individuality of the object, insofar as it is embedded in what he calls the "fabric of a tradition." That is, an authentic work of art is an object that was created at a certain time for a specific purpose. To sustain an aura of authenticity, a work of art has to have been involved in a sacred or quasi-sacred ritual function of some sort, such as in a magic or religious cult, or at the center of a community of worship. In secular cultures, the aura is preserved, in a slightly degenerate form, by what Benjamin calls "the cult of beauty," the secularized but quasi-religious worship of art for art's sake. (This is the reason why art galleries are like churches, with the works curated like holy relics: the point is to preserve their aura.)

So to qualify as an authentic work of art, it is essential that it be connected in some way to a community and its rituals, and the further removed an object is from this ritual power, the more the aura withers. This is why Benjamin thought that the early-twentieth-century debate over whether photography and film are legitimate forms of art completely missed the point. The real issue was the way in which these had completely transformed the entire nature of art by dissolving the relationships within which the concept of the authentic work made sense. The two main solvents at work in the age of mechanical reproduction are *massification* and *commodification*.

With these new kind of artworks, of which there can be any number of functionally identical copies, the question of which is the original ceases to make any sense. Once the work is cut loose from its place in the rituals of a community, indeed from the need to be in a specific place and time, we see the rise of the simultaneous collective experience – when a movie opens "in cinemas everywhere," everyone who sees it has the identical experience, across the city, even across the continent.

These new kinds of artworks also marked the transformation of art into a commodity, as it was pulled out of its primary role as part of a (quasi-) sacred ritual and turned (at best) into a vehicle for mass entertainment. At worst, art as a commodity ceases to be valued for its essential place in a living tradition and is turned into kitsch. This is the world of airport gift shops and tourist traps, of "authentic" African masks or Inuit soapstone carvings, the Disneyfied paintings of Thomas Kinkade or the rural sentimentality of Andrew Wyeth. Forget the aura; this is the stuff that barely registers on the consciousness as "art" at all.

By opening the door to art as a mass commodity, the age of mechanical reproduction created a crisis of authenticity in art. In the age of secularized, commercialized, mass-marketed entertainment, what plays the role of the ritual in preserving the aura of the work is the artist's life. Their past, their history, their lifestyle or persona is what provides the ballast that anchors the work in some sort of creative tradition or narrative, saving it from the frothy superficiality of mere commerce. That is why it matters more whether Avril Lavigne was a real skate punk than whether she wrote her own songs, and why we remain fascinated with the work of Andy Warhol, despite the fact that his whole artistic agenda was to blur the lines between commodities and art works. His life itself was a work of art, and when we buy a Warhol or apprehend one in a gallery, it is the aura of that life that we are appreciating.

One logical endpoint of this takes us to the world of contemporary art, where many of the works in and of themselves are so ludicrous in concept and so inept in execution that the old philistine war cry "My child could do that" is an insult to untalented children everywhere. But this objection misses the point, which is that the work itself is totally irrelevant. What is being sold is the artist himself, his persona or, better, his brand. And no contemporary artist has a better brand than Damien Hirst.

Shortly before the Wall Street crash in the fall of 2008, Sotheby's of London auctioned off 223 of Hirst's works. When the gavel finally fell on the last lot, Hirst was $200 million to the good, a record haul for an auction devoted to a single artist. Not all in the art world were impressed. The great critic Robert Hughes wrote a magnificently sour piece for *The Guardian*, in which he declared that the auction's only remarkable aspect was that it revealed the huge gap between the prices for Hirst's work and his actual talent. He called Hirst a "pirate" whose only skill is his ability to bluff and flatter the dumb, ignorant, and rich, and blamed him for almost single-handedly creating the cult of artist-as-celebrity. "The idea that there is some special magic attached to Hirst's work that shoves it into the multi-million-pound realm is ludicrous," he wrote.

But there *is* a special magic attached to Hirst's work. That magic is the spectacularly successful brand known as Damien Hirst. And for those to whom the brand is successfully marketed – hedge fund types, tycoons of all sorts, generally anyone else who happens to be cash-rich but taste-poor – it makes his products worth every cent.

In his book *The $12 Million Stuffed Shark*, business professor Don Thompson observes that "there is almost nothing you can buy for £1 million that will generate as much status and recognition as a branded work of contemporary art." As he says, some people think a Lamborghini is vulgar, and lots of people can afford yachts. But put a Damien Hirst dot painting on your wall and the reaction is, "Wow, isn't that a Hirst?" The point is, Hirst is not selling art, he's selling a cure for rich people with severe status anxiety. Hughes says of the shark, "One might as well get excited about seeing a dead halibut on a slab in Harrods' food hall." But snarkiness over sharkiness isn't serious art criticism, and judging Hirst's work by the criteria of technical skill, artistic vision, and emotional resonance is like complaining that the Nike swoosh is just a check mark.

The descent into the inanities of contemporary art is one natural consequence of the crisis of authenticity caused by mass reproduction of art, and it isn't even obvious that this is the sort of result that would have bothered Walter Benjamin. He was certainly wary of how the mechanical reproduction of art pushed it into the service of mass and even totalitarian politics. At the same time, he saw that widespread access to art had a democratizing influence, taking its consumption and critical appreciation out of the hands of the power brokers and the elites.

Yet through it all, Benjamin was fully aware of how the production of art remained, for the most part, in the hands of the elites in no small measure because the new technologies of mass art, photography and film in particular, were expensive and technically sophisticated. The next revolution would not occur until artistic production itself was democratized, rendered cheap, accessible, and instantly transmissible, in the age of digital reproduction.

My iPod is packed with thousands of songs I've never listened to by bands whose names I don't recognize. The hard drive of my laptop contains dozens of movies I've downloaded and never watched, and if all goes according to the pattern, I will soon have a Kindle or similar reader full of books I'll never read by authors I don't appreciate. I'm far from alone in this: in the age of digital reproduction, we treat art as a commodity – cheap, ubiquitous, and disrespected.

The old cyberlibertarian slogan declared, "Information wants to be free," but of course information doesn't want to be anything. It is just a good like any other, subject to the usual laws of supply and demand. For centuries information was scarce, and the heavy demand for news, culture, art, and other "idea-laden" goods made them expensive. We now live in a topsy-turvy world of information abundance, where a glut of ideas is chasing an increasingly limited supply of demand, in the form of time or attention.

There has been a lot of talk recently about the rise of the "freeconomy." This is a world where the marginal cost of producing another unit of culture – a song, a news story, a video – is approaching zero. This is the online digital economy that has been wreaking havoc with the business models of newspapers, magazines, and other enterprises that make a living by selling stuff made of ideas, now that those ideas can be copied at a marginal cost only a shade above zero. But one issue that has been somewhat neglected in that discussion is the effect of "free" on art itself, on the nature of aesthetic experience when the only expense is the time it takes to consume it.

In contrast with Walter Benjamin's era, which saw the mass consumption of art that remained centrally produced, in the age of digital culture it is not just access to art that has been democratized, but its production as well. What we are seeing now is the fulfillment of the Rousseauian ideal of every individual as a creative spirit, as millions of amateurs flood the Internet with their own songs, videos, photographs, and stories. But when everyone is so busy creating, who has time to consume any of it? In an economy where what is scarce is attention, the spoils will go to the artist who is best able to command it, even if this requires some rather baroque or contrived setups to achieve. For example, when Moby released his latest album, he booked an entire spa for a day so that journalists could listen to his new album while getting a massage.

A more delightful example of the attention economy at work comes courtesy of a fan of indie folk hero Sufjan Stevens. In 2007, Stevens held a contest in which he awarded the rights to a new song, "The Lonely Man of Winter," to a New York theater director named Alec Duffy. While Stevens gave him the unconditional right to do whatever he wanted with the song (destroy it, or use it to sell snowmobiles), most fans expected that Duffy would just put it online for all to hear. Instead, Duffy decided that the only place anyone would hear the song would be in his living room. Sufjan Stevens fans now make pilgrimages to Duffy's Brooklyn apartment, where he serves tea, plays the song a few times, and then sends them on their way with a bag of cookies, a tune they'll never hear again already fading in their minds.

Can you see what is happening here? It is the return of the aura, of the unique and irreproducible artistic work. Across the artistic spectrum, we are starting to see a turn toward forms of aesthetic experience and production that by their nature can't be digitized and thrown into the maw of the freeconomy. One aspect of this is the cultivation of deliberate scarcity, which is what Alec Duffy is doing with his listening sessions. Another is the recent hipster trend to treat the city as a playground – involving staged pillow fights in the financial district, silent raves on subways, or games of kick the can that span entire neighborhoods. This fascination with

FOUR

CONSPICUOUS AUTHENTICITY

HERE IS A SHORT BUT SOMEWHAT REPRESENTATIVE LIST OF brands, people, products, or services that have been marketed or promoted in recent years on the grounds that they are authentic: Italian cuisine, Chinese cuisine, Ethiopian cuisine, American cuisine, Canadian cuisine, Coca-Cola, Bailey's Irish Cream, distressed jeans, distressed guitars, skateboards, skateboarding shoes, books, independent bookstores, typewriters, chainsaws, Twitter, crowdsourcing, blogs, comments on blogs, ecotourism, communist tourism, slum tourism, Al Gore, John McCain, Sarah Palin, Barack Obama, Susan Boyle, Michael Phelps's mom, the Mini Cooper, the Volkswagen Beetle, botox, baseball, Samuel Adams beer, Russian vodka, English gin, French wine, Cuban chocolate, Cuba, Bhutan, organic coffee, organic produce, locally grown produce, locally grown organic produce, the 100-mile diet, the 100-mile suit, urban lofts, urban lofts with no-flush toilets, and mud floors in suburban homes.

We could add to this list indefinitely, to the point where this chapter would consist entirely of uses of the terms *authentic* or *authenticity* in selling things. The quest for authenticity is the contemporary advertising equivalent of the search for the holy grail, the ultimate marketing position that can elevate a brand above the "shiny, fabricated world of spun messages and concocted experiences." Being able to play the authenticity game is now a fundamental requirement of marketing, the standard against which all brand strategies are judged. Consumers are increasingly knowledgeable and sophisticated, and while the constant barrage of ads has made them wary, they are willing to go along with a brand that promises (and delivers) originality, integrity, and value. Calling something authentic, though, doesn't necessarily make it so, and the question of just what counts as authentic, and why, is one of the most pressing questions facing both producers and consumers.

When I was in high school in the 1980s, the only socially acceptable brand of jeans was Levi's. More specifically, you had to wear Levi's 501 button-fly red-tab jeans, not the cheaper orange-tab zipper-fly models, and most definitely not any of the competing brands such as Wrangler or Lee. Even the girls at my school wore the men's 501s, often a worn and faded pair they had borrowed (or stolen) from a boyfriend. Not only were men's 501s far more flattering on women than the high-waisted "mom jeans" that were in style at the time (anticipating by two decades the low-rise jeans phenomenon of the early 2000s, though it has to be noted that mom jeans briefly returned, thanks to the efforts of Jessica Simpson and Fergie), but they also allowed women to participate in the classic American style that went along with wearing a beat-up pair of button-fly Levi's.

The Levi's story is pretty well known, as retail narratives go. In 1873, a Bavarian immigrant named Levi Strauss invented riveted denim pants for Gold Rush laborers. The technique and the style caught on, and for most of the next century and a half Levi's jeans were true icons of working-class authenticity. Everyone who wore jeans wore Levi's, including James Dean, Marilyn Monroe, and Andy Warhol, not to mention every one of the two thousand students at my high school. By the mid-1990s, the company had sales of more than US$7 billion and owned almost 19 per cent of the

U.S. jeans market. Things were even better in Canada, where in 1991 its market share was 31 per cent.

But the accelerating rate of change in fashion trends and the changing structure of the retail market eventually caught up with Levi Strauss. It completely missed out on the hip-hop-inspired baggy-pants craze of the mid-1990s and then suddenly found itself fighting on two fronts, caught between low-cost retailers such as Walmart and Old Navy on the one side and the new high-end designer labels such as Juicy Couture and Seven on the other. By 2002, sales had dropped to more than $4 billion, and the company's share of the U.S. market was cut by a third, to 12 per cent.

Levi Strauss responded by chasing both ends of the barbell at the same time. For the low-end market, it entered into an agreement with Walmart to sell a low-priced "signature" line of jeans. Meanwhile, it decided to re-establish Levi's as the quintessential American brand by launching a line of "vintage" jeans that would be marketed based on claims about the historical authenticity of the various styles. In order to give the jeans their "historical" character, the company pays its workers to bleach, tear, scrub, acid wash, beat, re-sew, and patch its denim pants. According to the Levi Strauss Vintage Clothing website, "Classic styling, a focus on authenticity, and extreme attention to detail mean each piece of Levi's Vintage clothing is a mirror of its historical original." Once the pants are completely distressed and restored, those "vintage" Levi's sell for enormous prices. For instance, the 1890 501 Cinchback jeans – meant to replicate pants that railroad workers wore 120 years ago – sell for $350. Similarly, the 1967 Beat-up with happy repair jeans (with three patches in the rear and torn around the cuffs) sell for $275.

It isn't hard to follow the corporate reasoning here. Since the company's original success resided in its status as an authentic brand, the solution to its decline was to relaunch the brand, making "authenticity" an explicit selling point. It hasn't succeeded though, and you can take your pick of reasons. To begin with, just about every jeans manufacturer now sells pants that have been stone-washed, distressed, weathered, run over by monster trucks, thrown in a dumpster, slashed by machetes, or otherwise artificially aged to make it look like the wearer has just come from the coal mine, or maybe the chain gang. The absolute champ of working-class denim authenticity is the brand Prps, whose jeans are made in Japan on vintage looms from the 1960s, using organically grown cotton sourced from Africa. The old looms waste a lot of fabric, but as the label's owner explains, it is inspired "by the denim worn by actual workers before jeans became middle-class leisure wear." He also claims that the telling "flaws" in Prps jeans are more authentic than those found in cheaper hand-distressed brands because they are based on the wear-patterns found on pants worn by real drag-racing mechanics. The jeans retail for anywhere from $275 to $400 a pair.

More to the point, there is a world of difference between something actually being authentic and merely being called "authentic," and somewhere along the line Levi Strauss managed to trade the first for the second, and the brand has suffered as a result. But hang on: James Dean was no gold-rush laborer, and there is no evidence that Andy Warhol ever lifted anything heavier than a camera in his entire life. So why were Levi's more authentic when they were successful and being worn by the entire population of Glebe Collegiate Institute in 1987, as opposed to when they became usurped by FUBU, Seven, Prps, and Acne, and the company resorted to relaunching the brand by highlighting its authentic heritage?

The answer appears to revolve around a distinction between the "genuine" authentic and the "fake" authentic. It is a tremendously important distinction, one that manifests itself in popular attitudes toward music, art, and literature as well as in our choices about where to go on vacation, whom to vote for in elections, and, yes, what jeans to buy. What that difference consists in, and how we even know what it looks like, is very difficult to say – though what corporations *do* know is that it is a difference that is worth a great deal of money. But first we need to address something even more fundamental, namely, how was authenticity transformed from a quest for individual freedom and self-fulfillment into a marketing strategy in the first place? The beginning of the answer takes us back to some familiar names in a familiar place: mid-eighteenth-century Paris.

Denis Diderot was born in 1713 in the small French town of Langres. In school, he began studying for the priesthood, first with the Jesuits and then under Jansenist supervision, but he abandoned the clergy when he was nineteen. He briefly studied law, decided to become a writer, and then spent most of his twenties as a free-thinking bohemian, living off of friends and the proceeds of sporadic tutoring work. It was in one of the cafés of Paris that he hooked up with the ambitious young Jean-Jacques Rousseau, and the two of them formed a powerful friendship, spending a great deal of time strolling in the Jardin du Luxembourg talking about music and literature and how to get ahead in the world.

Even if he were known only for his work as a novelist, playwright, and art and social critic, Diderot's place in history would be secure. His great claim to fame, though, is as the driving force behind the *Encyclopédie*, a mammoth and politically subversive publishing venture that brought together some of the greatest minds in Europe, including Jean d'Alembert, Voltaire, and Rousseau himself. Even as he devoted his tremendous energies to steering the *Encyclopédie* through treacherous political and financial waters, Diderot found time to write a number of satirical literary works, on which he worked in secret to avoid endangering his day job, and which were only published after he died.

The most famous of these posthumous works is *Le Neveu de Rameau* (*Rameau's Nephew*), which Diderot drafted in 1761 and spent the next decade and a half polishing. It was included in the pile of papers and books that Diderot left to his patroness, Catherine the Great, upon his death in 1784. It would have remained buried there were it not for an illicit copy that was smuggled into Germany in 1803. The manuscript reached poet and philosopher Friedrich Schiller, who passed it on to Johann Wolfgang von Goethe, who fell so in love with the work that he undertook the job of translating it. The German version became a sensation, and it was read with admiration by Karl Marx, Sigmund Freud, and especially the philosopher Georg Wilhelm Friedrich Hegel, who considered it to be a work of "exceptional significance, the paradigm of the modern cultural and spiritual situation."

For all the high-powered praise, and for all its enduring fame, *Rameau's Nephew* is not an easy work to follow, let alone summarize. It is basically a long dialogue between a character named "I" (who is assumed to be a stand-in for Diderot himself) and a character called "He," who is the unnamed nephew of the famous composer Jean-Philippe Rameau (the same Rameau we met in chapter 2, who told Rousseau that he had no future in the music business). The nephew is a sponger and a wastrel who spends most of his time sponging off insipid rich acquaintances, who are happy to give him supper as long as he laughs at their jokes and pretends to share their bourgeois principles. The "I" character (who for simplicity's sake I'll call "Diderot") is a more straightforward fellow, a social conformist committed to the noble virtues of truth and honesty.

The dialogue is a wrangling sort of conversation, meandering from the meaning of morality and the nature of genius to the proper way to educate children, but along the way the two cheerfully gossip about dozens of their contemporaries, especially famous and semi-famous musicians, artists, journalists, and public figures. There is a lot of debate among scholars over just what Diderot's intention was in writing the piece, and to some extent the confusion arises because the author's sympathies do not completely lie with either the Diderot or the nephew figure. There are multiple layers of irony coursing through the work, with each character taking turns contradicting himself or making absurd statements. This is especially true of the nephew, whom Diderot describes, in a bit of a stage whisper, as "a compound of elevation and abjectness, of good sense and lunacy" who has "no greater opposite than himself."

In this piece we find the earliest pointers of the direction that the culture of authenticity was going to take over the course of the nineteenth century. In his introduction to his own translation of the work, the historian of ideas Jacques Barzun claims that in *Rameau's Nephew* we find the eloquent first statement of the ideas that were to define the reaction to modernism in the nineteenth century, including Romanticism, the worship of genius and creativity, and the emphasis on self-discovery. Barzun is echoing Hegel, who saw the piece as a terrific indictment of society, of its thoroughgoing falseness, and of the base insincerity and hypocrisy of everyday life. It is from this society that the nephew – a parasitic sensualist who is bitter at the realization that he is not a genius – is alienated, systematically separated from his actual self.

The nephew character spends a great deal of time complaining that in order to advance in the world (or even just to eat), he is forced to spend his time assuming various "positions" or social roles, performing what he calls "the beggar's pantomime." Yet when it comes to finding one's place in society, no one, not even the king himself, is free of the obligations of the beggar's pantomime. As Diderot lectures the nephew:

> Whoever stands in need of another is needy and takes a position. The King takes a position before his mistress and before God . . . the minister trips it too, as courtier, flatterer, footman and beggar before his king. The crowd of self-seekers dance all your positions in a hundred ways, each viler than the next . . .

This is not so much an echo as it is a direct restatement of Rousseau's critique of civilization, wherein the problem is society itself. Because we are not self-sufficient, because we need things from one another, we are forced to adopt various poses as we jockey for favors and for status. What sets the nephew apart from all the other players in the beggar's pantomime is that he, at least, sees it for the humiliating act it is. He is able to keep his social roles at arm's length, taking an ironic stance that at least prevents his identity from being subsumed by the mask he wears to the pantomime. The price of this ironic distancing is alienation, because the nephew must continue to struggle to get by in a social milieu in which he can never be true to himself. But it is not clear that he has any serious alternatives. As even Diderot concedes, the one human in history who has managed to avoid being forced into one "position" or another is the Greek philosopher Diogenes the Cynic, who "made fun of his wants," went naked, slept where he fell, and ate whatever he could scrape together.

From this description, it is clear that Diderot has it wrong: there have been plenty of people throughout history who have managed to escape playing a part in the beggar's pantomime. We call them cranks, eccentrics, hobos, homeless people, misfits, and any number of other terms to describe their resolute failure to fit in. Sometimes (as in the case of Diogenes) the pose is deliberate, but sometimes it just speaks to a deep inability to cope with the social demands of everyday life.

At any rate, it is one thing to complain about society, another entirely to choose to reject everything society has to offer in the way of subsistence, not to mention comforts. The nephew objects that going without food, shelter, or clothing is too high a price to pay for happiness, that it is better to play a part in the pantomime than to "crawl, eat dirt, and prostitute yourself." Diderot responds by calling him a greedy lout and a coward, but it is hard for us, as readers, to take the reprimand seriously. After all, if he accepts the nephew's critique of society as sound, then the comfortable and conformist Diderot is just as greedy and just as cowardly. Like Socrates, who claimed that what separated him from his ignorant fellow Athenians was that at least he, Socrates, knew that he knew nothing, Rameau's nephew is one up on his fellow Parisians insofar as he is at least willing to acknowledge that he is a sellout, wallowing in his alienation.

That is why the protagonist of *Rameau's Nephew* has been held up for over two centuries as a sort of antihero of authenticity, the principled (if bitter) antisocial nonconformist who entirely rejects the values of bourgeois society, and whose legions of literary descendants include Charles Baudelaire, Thoreau and Emerson, the Beat poets of the mid-twentieth century, and all of the countless post-Romantic heroes of fiction, music, television, and film, from Madame Bovary and Holden Caulfield to Lenny Bruce and Kurt Cobain.

If we have trouble following Hegel in seeing the nephew as the avatar of a new individual consciousness, one that "looks upon the authoritative power of the state as a chain . . . obeys only with secret malice and stands ever ready to burst out in rebellion," the familiarity and the ubiquity of the pose may be precisely the problem. The idea that authority is repressive, that status-seeking is humiliating, that work is alienating, that conformity is a form of death . . . none of this is remotely original. We have heard every variation of the tune, from nineteenth-century bohemians to twentieth-century counterculturalists to twenty-first-century antiglobalists, and we know every part by heart.

It is not the sheer persistence but rather the amazing popularity of the stance that ought to give us reason to pause and maybe reconsider our attitude toward modernity. Look around. Is there anyone out there who does *not* consider him or herself to be an "antihero of authenticity"? Anyone who embraces authority, delights in status-seeking, loves work, and strives for conformity? Sure, there are a few, we even have names for them. We call them drones, widgets, squares, yuppies, fascists, but nobody ever admits to being a drone or a yuppie or a widget. Living inauthentically is always something *other* people do. In which case, what is surprising is just how much apparent inauthenticity there is out there.

In an essay entitled "Fake Authenticity: An Introduction," Boston-based writer Joshua Glenn argues that the character of Rameau's nephew, with his cultivated alienation and easy nonconformity, is far from the first antihero of the ethic of authenticity that Hegel and others imagined him to be. Rereading *Rameau's Nephew* in the light cast by writers such as Norwegian-born American economist Thorstein Veblen and his contemporary disciple Thomas Frank (whose 1997 book *The Conquest of Cool* was the first systematic exploration of the marketing of cool), Glenn says we can now appreciate the nephew as the first hero of *fake* authenticity, the striving face of hip capitalism, and "an architect of consumer dissatisfaction and of perpetual obsolescence."

By "fake authenticity," Glenn doesn't mean that which is simply fake, but something more sinister. As a first pass at making this distinction clear, we can say that the difference consists in the following: to call something fake implies that it corresponds to something real. In contrast, the fake-authentic does not imply something genuine of which the fake is a mere facsimile. So the agenda of fake authenticity then is to replace the whole real-versus-fake game with one where that distinction no longer serves any purpose, or even makes any sense.

An example may help make it clearer. In Nashville, Tennessee, not far from the campus of Vanderbilt University, there is a full-sized reproduction of the Parthenon. Built in 1897 as the centerpiece of the Tennessee centennial fair, it was originally made of wood, brick, and plaster, but it was rebuilt in 1920 out of concrete. For all its impressiveness, the Nashville Parthenon is a fake, an acknowledged copy or replica of the real Parthenon that remains where it has always stood, in Athens, Greece. The key word here is *acknowledged*, since the point of building the replica was to pay homage to Nashville's status as "the Athens of the South." The existence of the fake Nashville Parthenon involves an explicit reference to the Greek original.

The point of the fake-authentic, in contrast, is to make itself appear as the genuine or original. Examples abound, although perhaps the most common (and unpleasant) is the way the oldest parts of virtually every North American city have congealed into tourist-friendly versions of their original selves. Or as Joshua Glenn writes about Boston, everything that was actually old "has been made Olde instead; historical façades and interiors have been restored not to how they used to look, but to how (city planners imagine) tourists want them to look." The point is, when a city's old center is made over into a touristified Olde Town, what remains is not a replica of things as they used to be, but what Jean Baudrillard calls the "simulacrum": not a copy of the real, but something that becomes "true" in its own right while bearing no relation to any reality, past or present, whatsoever.

A more recent exposition of the fake/fake-authentic distinction is found in the philosopher Harry Frankfurt's best-selling pamphlet from 2005, "On Bullshit," which explores the difference between a lie and bullshit. As Frankfurt argues, there are two elements to a lie. First, the liar must believe that his statement is false; second, his intent in making that statement must be to deceive his interlocutor about the facts. Thus, the liar is playing the truth-telling game, insofar as it is concerned with the relationship between a statement and the world. So just as the fake Parthenon in Nashville acknowledges the existence of the real Parthenon in Greece, the fact that a statement can be outed as a lie at least acknowledges the existence of its opposite, the truth.

Bullshit makes no such acknowledgment. What characterizes bullshit is a complete lack of concern for both truth and falsehood – bullshit isn't even in the game of lying or telling the truth. That is why the cardinal virtue of bullshit is not correctness, but *sincerity*: "Rather than seeking primarily to arrive at accurate representations of a common world," writes Frankfurt, the bullshitter "turns toward trying to provide honest representations of himself." Therefore, it is hardly surprising to find that the two areas of human enterprise most concerned with sincerity *as opposed to the truth* – namely, politics and advertising – are also the two areas most steeped in bullshit. Or would it be better to say that politics and advertising are the two areas most concerned with the appearance of authenticity? This might be a distinction without a difference.

We have finally wormed our way to the question we have been circling around until now, which is whether authenticity even exists. Is there a genuine authenticity that stands to fake authenticity, as the true stands to the false? Joshua Glenn, following Baudrillard, says the answer is no, that whenever you find something described as authentic, you know that you are already in the realm of fake authenticity. He writes, "Although Italians do open restaurants, there is no such thing as an authentic Italian restaurant. Although history, nature, race, and class are very real and very much with us, there is no such thing as an authentic past, an authentic outdoors, nor an authentic non-white/middle-class style of life."

The problem Glenn is getting at is twofold. First, he is pointing out the way self-consciousness about authenticity is self-defeating. Authenticity is like authority or charisma: if you have to tell people you have it, then you probably don't. The second, related point is that authenticity has an uneasy relationship with the market economy. This is because authenticity is supposed to be something that is spontaneous, natural, innocent, and "unspun," and for most people, the cash nexus is none of these. Markets are the very definition of that which is planned, fake, calculating, and marketed. That is, selling authenticity is another way of making it self-conscious, which is again, self-defeating.

What Joshua Glenn is presenting is another variation on the classic co-optation narrative, and it goes like this: Once upon a time, there were authentic miners wearing jeans or authentic Italians running restaurants, there were old parts of town with authentic industrial lofts or cobblestone streets, or an authentic subculture such as punk, with its own scene, look, sound, and way of speaking. Then one day the capitalists looked over and saw how appealing all this authenticity was and thought that maybe people who weren't miners (or Italian or industrial-loft-dwellers or punks) might like to partake in that authenticity. And so they figured out a way of taking the signs and symbols of that authenticity, draining the authentic credibility out of it, and selling a fake-authentic version to the credulous masses. In this view, authenticity really is nothing more than a marketing ploy.

It is convenient, and for many people very satisfying, to blame the capitalists, but it isn't that simple. In fact, the problem with authenticity is not that it doesn't exist, or that it is nothing more than an advertising slogan, a cheesy shell game, rigged by the capitalists to sell jeans and beer and exotic vacation packages. It is true, authenticity is a moving target, and as soon as we think we have a lock on it, it has danced away. But what is really driving the quicksilver character of the search for authenticity is the underlying competitive structure of the quest. That is, we should not blame those who are selling the authentic, but rather those who are buying.

Around the world, hundreds of millions of people live, work, and sleep on mud floors. Not out of choice, mind you, but because they're literally dirt-poor. If they had a choice in the matter, they'd probably prefer to have a floor made out of wood, or maybe marble if they're really ambitious. In some parts of the United States though, having a mud floor in your house is not a sign of desperate poverty but of your impeccable authenticity.

The trend toward "earthen" (which is a fancy word for *mud*) floors is part of a "natural building" movement that eschews unnatural flooring such as linoleum and even rejects unnaturally sawn wood planks or unnaturally squared ceramic tiles. Environmentally conscious people want to use materials that have as little "processing" as possible – so they look for floors made of straw, bamboo, and dirt. As one couple profiled in *The New York Times* put it, mud flooring helped turn their house into a "sacred space" – they even mixed the blood of an ox into the mud before spreading it on their floor, to give it an authentically spiritual aspect. "I think people are craving the earth," the owner told the *Times*. "They want to be more primal. How much more primal can you get than dirt?"

It will perhaps come as no surprise that this trend is most evident on the West Coast, which has for decades been on the cutting edge of the avant-authentic. It is also a trend that would have made immediate sense to Veblen, whose 1899 book *The Theory of the Leisure Class* did for our understanding of capitalism what Darwin's *On the Origin of Species* did for biology.

That is not raw hyperbole. Just as Darwin remade biology by completely upending the ancient Aristotelian theory of essential biological types, Veblen's work remains a vital antidote to the Marxist theory of capitalism as a top-down system of exploitation of the proletariat by the bourgeoisie. For Veblen, any adequate account of the driving motors of a capitalist economy has to pay special attention to the role of status-seeking, in particular to the various forms of conspicuous consumption designed to evoke what he calls an "invidious comparison" – a comparison that judges people according to their relative worth or merit and is designed to evoke envy or resentment.

Veblen's account of the goals and motivations behind human behavior is remarkably similar to that of Rousseau. According to Veblen, there are two basic "instincts" or "proclivities" that govern human action: the "instinct of workmanship" and the "predatory

proclivity." The instinct of workmanship is, essentially, an appreciation for useful and efficient work, and it arises directly out of the
nature of most work as directed at a given survival-enhancing goal.
Given the need to hunt and fish, gather food and fuel, build shelter,
and so on, we naturally come to value work that is done skillfully
and efficiently, while disparaging work that is shoddy or ineffective.
In contrast, the predatory proclivity manifests itself in social situ-
ations, through fighting or other forms of dominant behavior.

These two instincts map fairly closely on to the Rousseauian
distinction between the natural and self-directed instinct for survival that he calls *amour de soi*, and the *amour-propre* that involves
someone establishing his or her self-worth through competition
and comparison with others. Furthermore, both Veblen and
Rousseau note that the instinct for workmanship (or self-love) can
result in a form of natural inequality, since some people are simply
better at hunting or starting a fire than others. The crucial step into
a predatory culture is taken when this natural inequality between
individuals becomes a springboard for unequal social relations
based on fear and dominance. That is, what was already a mere
ranking (some men are better hunters than others) becomes a
dominance hierarchy, where men take pride in exploiting the
weaknesses of others. In a sentence that Rousseau might have
written, Veblen remarks that the transition to a predatory culture
is, essentially, the transition to civilization, "from a struggle of the
group against a non-human environment to a struggle against a
human environment."

Once a society becomes rich enough to sustain an economic
surplus, its predatory character takes the form of a class-based
society. Because it is able to sustain itself by appropriating for itself
a large share of the economic surplus, the upper class becomes a
"leisure class," that is, exempt (or, in many cases, forbidden) from
engaging in useful or productive work. The name is not really accurate though, since members of the leisure class do not typically

spend their time lounging around. Veblen argues that even in a predatory culture, the instinct of workmanship remains in force. That is why the members of most leisure classes feel that it is necessary to spend their time and energy on activities that have a veneer of the useful, but which clearly – if subtly – demarcate the performer as a member of the upper class.

The best way of looking busy without actually doing any productive work is to engage in activities that were once quite useful but that are now vestigial. In a society that is just past the early stages of modernization, these usually take the form of pre-industrial forms of aristocratic exploit, preferably with a warlike edge to them. That is why the upper classes in the nineteenth century engaged in hunting, sports, and swordplay while becoming learned in quasi-diplomatic spheres such as aesthetic appreciation, etiquette, and esoteric learning of complicated or obsolete languages.

The irony is that all of this useless display of skill takes a lot of effort, to the point where members of the leisure class have always found themselves working at least as much as, if not more than, they would need to as members of the productive classes. This is a pattern that remains in place today, although the nature of the useless exploit has changed significantly. To take just one example: in 2006, the hotel heiress, Internet sex-tape star, and drunk-driving celebutante Paris Hilton was raked over the coals for an interview she gave to *Hello! Magazine* in which she described herself as, inter alia, a brand, model, actress, designer, and artiste, and said, "I worked hard for all this. I tell girls that if you basically work hard, all your dreams will come true." The fact is, Paris Hilton *does* work hard. Yes, she gets paid half a million dollars to do pointless things, like fly to Austria to host a party, but that these activities are pointless and lucrative does not alter that fact that they are also quite exhausting.

Veblen's reputation today suffers from a couple of misunderstandings. First, his account of the leisure class is frequently assumed to apply only to the late-nineteenth-century American pseudo-aristocracy. At the same time, his work was picked up by midcentury "progressive" intellectuals such as Lewis Mumford and John Kenneth Galbraith, who tended to read Veblen as a prescient and moralizing critic of the new consumer society.

Neither of these interpretations is accurate. It is true that Veblen condemned the activities of the leisure class for being "wasteful," but he meant it in a rather technical sense. He didn't think that swordfighting or the study of ancient Greek are wasteful in themselves; on the contrary, both activities do a fine job of feeding the demands of the instinct of workmanship. Their wastefulness lies in the status competition that emerges as a result of the invidious nature of these exploits. If knowledge of ancient Greek is a sign that one is a member of the leisure class, then surely it follows that knowing both ancient Greek *and* Latin signals that one has even more time to spend on esoteric learning, and is thus a source of even higher status.

All forms of status competition are zero-sum games, since in order for one to gain in status, someone else must lose. It is easy to see, then, how leisure class exploits will quickly take on the character of a classic arms race, with more and more effort being spent by everyone involved just to maintain their same relative place in the pecking order. It is this collective expenditure of resources that Veblen saw as wasteful, even though it is perfectly rational behavior from the perspective of each member of the leisure class. Status has real value after all, and people cannot be faulted for pursuing and defending it.

When people read Veblen today, they commonly look at his specific examples of conspicuous consumption and conclude that they are exempt from Veblen's critique, since they don't engage in these sorts of obvious and frequently ridiculous displays of wealth or status. What we need to always keep in mind is that it is precisely because they are so obviously concerned only with status that we find various forms of quasi-aristocratic posturing so easy to mock. Long before he was convicted of mail fraud by a Chicago jury, embattled newspaper tycoon Conrad Black had been tried and convicted by his fellow Canadians for the far more serious crime of being a crass social climber. What made it an open-and-shut case was the obsolete nature of Black's efforts. Already well known for an arcane vocabulary that made him sound like he had been raised in India under the Raj circa 1839, as well as for a hobby that involved moving figurines representing Napoleon's armies around on large-scale maps, Black sent his countrymen into fits of laughter when he surrendered his citizenship in order to take up an ermine-clad position in the British House of Lords, perhaps the only parliamentary institution in the world that makes Canada's own appointed-for-life Senate look vigorous and spry in comparison.

Black's considerable efforts at conspicuous display were undermined by the sheer obviousness of his intentions. If only he had read Veblen, Black might have understood that what made him a laughing stock among mainstream society and a bit of an embarrassment to his intended peers was that, in engaging in such conscious status-seeking, he was inadvertently exposing the artifice at the heart of the leisure class's pretensions. Remember, it is essential to the pretences of the upper classes that their activities of conspicuous leisure be at least superficially useful. Their real function – to demarcate the social classes – must be left implicit, to be decoded only by those who are truly in the know. This is why, as Veblen predicted, the behaviors and activities that constitute upper class "waste" have evolved in lockstep with the patterns of consumption and employment in society as a whole. As it turns out, Conrad Black's mistake was not that he was a status-seeker, it was that he chose a rather unfashionable ladder to climb.

When most people think of status, they think of the rigid class structures of old Europe. In contrast, North America is considered to be a relatively classless society. Sure, we have various forms of

inequality, income being the most obvious and most socially pernicious, but we have no entrenched class structure, no aristocracy
that enjoys its privileges explicitly by virtue of birth, not merit.
Nevertheless, urban North Americans live in what is probably the
most status-conscious culture on the face of the Earth. The reason
we don't recognize this fact is that most of us are stuck in a model
derived from the old aristo/bourgeois/prole hierarchy, where status
is linear and vertical, a ladder on which one may (or may not,
depending on the status markers that are in play) be able to move
either up or down.

That model of status is pretty much obsolete. Over the course of
the twentieth century, the dominant North American leisure class
underwent three distinct changes, each marked by shifts in the relevant status symbols, rules for display, and advancement strategies.
The first change was from the quasi-aristocratic conspicuous
leisure of Veblen's time to the bourgeois conspicuous consumption that marked the growing affluence of the first half of the twentieth century, a pattern of status competition commonly referred to
as "keeping up with the Joneses." The next change was from bourgeois consumerism to a stance of cultivated nonconformity that
is variously known as "cool," "hip," or "alternative." This form of
status-seeking emerged out of the critique of mass society as it was
picked up by the 1960s counterculture, and as it became the dominant status system of urban life, we saw the emergence of what
we can call "rebel" or "hip" consumerism. The rebel consumer
goes to great lengths to show that he is not a dupe of advertising,
that he does not follow the crowd, expressing his politics and his
individuality through the consumption of products that have a
rebellious or out-of-the-mainstream image – underground bands,
hip-hop fashions, skateboarding shoes.

In both cases, what motivated the switch from one status system
to another was a combination of economic, social, and political evolution that made the old status moves too outdated, too obviously

useless. The shift from aristocratic leisure pursuits to a bourgeois form of conspicuous consumption was a consequence of the consumer revolution of the first half of the twentieth century and the growing industrialization of the economy. At that point, it was no longer possible to even pretend that fox hunting or learning Latin served any remotely useful purpose, and they became seen for the self-conscious exercises in status-seeking that they always had been. Similarly, keeping up with the Joneses was fine for a postwar society that valued hard work and material comfort, but it became increasingly intolerable as the culture, especially the world of the educated elites, became dominated by the anticonsumerist values of the counterculture.

For most of the past forty years, the values of educated urbanites (sometimes known as the "cultural elites") have been dominated by the politics of cool, a set of countercultural ideals that rejects conformity in all its guises and that puts a premium on a pattern of consumption that confirms one's status as a nonconforming rebel. Instead of keeping up with the Joneses, it was necessary to "get down with the Joneses," which involved a perpetual coolhunt. But thanks to incredible advances in communications technology, cool has ceased to be credible as a political stance.

When I was going to high school in the 1980s, I lived in Ottawa, Canada, a stuffy, boring, and bureaucratic city. The radio stations all played (and still play) only classic rock, and there was very little in the way of alternative culture. There was one record store in town that stocked the latest albums by bands such as the Cure, Siouxsie Sioux, the Jam, and U2. Every few months, the owners of the store would travel to New York or London to see new bands, to dig through record stores, and walk around looking at what people were doing, wearing, and listening to. This was an early form of coolhunting. It was also very slow.

This meant that a subculture – of art, or fashion, or music – could sit, isolated and protected, in a small enclave in New York, London, or Berlin. Camden Town or SoHo served as hothouses of cool, and it could take months, even years, before the new styles or sounds were noticed by the mainstream media and communicated to the masses. This gave the impression that there were two cultures, with two economies. There was the economy and culture of mass society, based on mass media and the mass production of homogenized goods, and there was the counterculture or subculture, which had its own distinct fashions, based around a do-it-yourself ethos and individualized and local forms of commerce. Because of this, it was possible to see the two cultural economies as having opposing interests and agendas. The mainstream economy was always trying to discover, copy, and co-opt the subversive styles of the counterculture, which in turn remained ever-vigilant and on the move. To be cool, in short, was to be a warrior on the front lines of resistance against the capitalism of mass culture.

Norman Mailer set the agenda in the 1950s when he wrote that society was divided into two types of people: the hip ("rebels") and the square ("conformists"). Cool (or hip, alternative, edgy) here becomes the universal stance of individualism, with the hipster as the resolute nonconformist refusing to bend before the homogenizing forces of mass society. In other words, the notion of cool only ever made sense as a foil to something else, that is, a culture dominated by mass media such as national television stations, wide-circulation magazines and newspapers, and commercial record labels. The hipster makes a political statement by rejecting mass society and its conformist agenda.

Like George Bush's infamous declaration that "you are either with us or you are with the terrorists," this Manichean characterization of cool gives it a great deal of power as a form of political and social criticism, by allowing us to situate everyone on one side or the other of a great divide. Either you are over here with the hip-sters, or you are over there with the conformist (and latently fascist) squares. But the truth is, cool is not political, and it never was. The reason why anyone ever thought that being cool did have political consequences was because of the tremendous amount of friction in the transmission of culture. It took a long time for subcultural trends in fashion or music or speech to move from the streets of London or New York City to the suburban basements of Omaha or Ottawa, and the phenomenon we call "cool" was just a consequence of that friction. Cool people were just those who had early access to new cultural trends, which gave them a great deal of status. Not only did it allow them to portray themselves as political radicals, it also allowed them to treat those who were not "in the know" as the mindless dupes of mass society.

Starting with the arrival of MTV in the early 1990s, technology has been remorselessly eliminating the time-lag between the birth of a cool subculture and its propagation throughout the main-stream culture. MTV itself has been replaced by the Internet and by websites such as MySpace or iTunes. The iPod has made the sharing of music around the globe a matter of pointing and click-ing. Cool can be transferred from Sweden to South Africa in a matter of minutes.

This means that there is no longer a mainstream versus a counterculture; there is only the "hipstream." The mass-media ecosystem has disappeared, and the prevailing aesthetic is not cool but quirky, dominated by unpredictable and idiosyncratic mashups of cultural elements that bear no meaningful relation-ship to one another. Appreciating the antilogic of quirk is the only way to navigate the movies of Wes Anderson (Jeff Goldblum in an "I'm a Pepper" T-shirt!) or the various tangents of Dave Eggers's McSweeney's publishing empire. For a daily dose of the quirk aes-thetic, go to www.boingboing.net, a "directory of wonderful things" that gets well over three hundred thousand visitors a day. A typical week of entries will draw your attention to a video of a man dropping forty-five pounds of Silly Putty off a building, an archive of Soviet-era children's cartoons, and a make-your-own-sex-toys blog. There's no rhyme or reason to any of it, apart from that it is all, in its own quirky way, kinda neat.

Cool fizzled out when it was exposed as just another consumerist status hierarchy, and when it passed so deeply and so self-consciously into the mainstream that it became simply embarrassing. Rebel consumerism became too obviously "wasteful" in the Veblenian sense once technological evolution combined with economic growth to make most "alternative" consumer goods available to just about anyone who wanted them, at exactly the same time. You like that Nice Collective jacket that Chris Martin is wearing in the new Coldplay video? You can buy the same one on eBay and have it delivered in the morning. Heard about a great new band that is playing a small club on the outskirts of town this weekend? Thanks to Web-alert services such as Flavorpill or Thrillist, so does everybody else. In the end, rebel consumerism died when it became a game that anyone with an Internet connection and a decent-paying job could play.

But status, like power, abhors a vacuum, and the thought that the end of cool might be the end of status-seeking is pure wishful thinking. Status-seeking never disappears – when it is exposed to the light, it simply scurries away and hides until it can transform itself into a subtler and less obvious form. Following the classic Veblenian pattern, those who are truly in the know have moved on from cool. The trick now is to subtly demonstrate that while you may have a job, a family, and a house full of stuff, you are not spiritually connected to any of it. What matters now is not just buying things, it is taking time for *you*, to create a life that is focused on your unique needs and that reflects your particular taste and sensibility.

Do you subscribe to an organic-vegetable delivery service? Do you believe that life is too short to drink anything but wine straight from the *terroir*? Do you fill your house with heirlooms, antiques, or objets d'art that can't be bought anywhere or at any price? For your next vacation, are you going to skip the commercialized parts of Europe or Asia and just rent yourself a cabin in British Columbia or a farmhouse in Portugal, away from all the tourists and the people trying to sell you stuff? Welcome to the competitive and highly lucrative world of conspicuous authenticity.

What makes conspicuous authenticity so seductive and appealing is the twist it puts on Veblen's insight that in order to be successful, the signs of conspicuous display need to portray themselves as at least superficially useful or socially beneficial. That is, it needs to masquerade as something other than what it really is, which is status-seeking. And so recall how the old nineteenth-century aristocrats spent their "leisure" time hunting or learning obscure languages, while late-twentieth-century counterculturalists masked their coolhunting under the guise of a principled rejection of fascistic conformity.

Conspicuous authenticity raises the stakes by turning the search for the authentic into a matter of utmost gravity: not only does it provide me with a meaningful life, but it is also good for society, the environment, even the entire planet. This basic fusion of the two ideals of the privately beneficial and the morally praiseworthy is the bait-and-switch at the heart of the authenticity hoax. This desire for the personal and the public to align explains why so much of what passes for authentic living has a do-gooder spin to it. Yet the essentially status-oriented nature of the activity always reveals itself eventually. A perfect case study is found in the rise and fall of organic produce.

As recently as a decade and a half ago, organic food (especially produce) was the almost exclusive bastion of earnest former hippies and young nature lovers – the sort of people who like to make their own granola, don't like to shave, and use rock crystals as a natural deodorant. But by the turn of the millennium, organic was making inroads into more mainstream precincts, driven by an increasing concern over globalization, the health effects of pesticide use, and the environmental impact of industrial farming. The shift to organic seemed the perfect alignment of private and public benefit.

It helped that a period of sustained economic growth had raised living standards almost across the board, making it possible for more people to afford to regularly spend two dollars for a pound of apples, or six dollars for a gallon of milk. As organic got more popular it moved into broader swaths of the economy, to the point where virtually any consumer good worth having now comes in an organic variant. And so there is not just organic produce, meat, and poultry, and baking, but also organic shoes, clothes, and dry cleaning.

The core product line though has always been food, especially produce. One of the great vehicles for bringing organic food to the masses is Whole Foods (known as "whole paycheck"), an enterprise that began as a natural foods supermarket in Austin, Texas, in 1980. It expanded into California in 1989, into Manhattan in 2001, Toronto a year later, and London in 2007. That same year it bought the 110 stores of its biggest rival, Wild Oats Markets, and by the end of 2009, it was operating almost 300 stores in the United States, Canada, and the United Kingdom.

Organic is now a thoroughly mainstream business. Almost every grocery store of a decent size has an organic section, and a 2007 poll found that almost one-third of Americans buy organic food at least occasionally. Yet for all its success, the case for organic has never been that solid, and the supposed benefits (the ones that justify the substantial price premium) have never been conclusively established. There is very little evidence that organic produce is better for you than its conventional counterpart, and the widely circulated claims in favor of the superior nutritional content of organic food are highly disputed. Also, the sustainability of organic farming itself is in question, given that nitrogen fertilizer produced from atmospheric nitrogen is not allowed under organic farming rules. One widely cited study estimates that if organic farming were to replace conventional farming worldwide, we would need the manure from another 7.8 billion cattle to replace the aerial nitrogen that conventional farmers currently use. Simply feeding that many cattle would require far more land than we have available.

When pushed on these issues, many advocates of organic eating fall back on a straightforwardly aesthetic claim: it simply tastes better. That may be so – after all, there's no accounting for taste, as Immanuel Kant observed long ago. But subjective reports of taste, appreciation, and satisfaction across the entire range of human experience are beset by a problem that psychologists and economists call "framing effects": the widely observed phenomenon whereby how we experience something is determined to a large extent by our expectations. One study, for example, gave two groups of people the same wine to sample. One group was told it was an extremely expensive wine of rare vintage, the other was told it was plonk. The first group reported much higher levels of appreciation for the wine than did the second. Similarly, for many people simply knowing they are eating organic food "frames" the experience a certain way, so they are psychologically primed to achieve a maximum satisfaction from the meal.

None of this has stopped organic from becoming a marketing phenomenon. It has also become an essential element of any "authentic" lifestyle; as *The New York Times* food writer Mark Bittman put it, for many people organic food has become "the magic cure-all, synonymous with eating well, healthfully, sanely, even ethically." Yet as it became more popular, the rumblings of discontent within the organic movement became harder to ignore. What was once a niche market had become mainstream, and with massification came the need for large-scale forms of production that, in many ways, are indistinguishable from the industrial-farming techniques that organic was supposed to replace.

Whole Foods took the brunt of public dissatisfaction, as customers rebelled against the increasingly "corporate" feel of the stores. The company was even called out by author Michael Pollan in *The Omnivore's Dilemma*, a book that has become the *No Logo* of the new foodism. He called Whole Foods hypocritical for playing up its sustainable-agriculture values while buying most of its produce from huge growers such as Earthbound Farm and Cal-Organic. Pollan accused Whole Foods of violating the local and small-scale spirit of the organic ethos, if not the exact letter of what constitutes organically grown food.

But what really sent organic advocates around the bend was when Walmart got into the game. In early 2006, the company announced that it was going to start offering a full range of organic products at only a 10 per cent premium over their conventional equivalents. If you honestly believe in the purported benefits of organic food, this could only be seen as a good thing. For the first time, millions of shoppers would have access to organic food at only a slight markup, leading to a healthier population and a better environment.

Except the reaction was almost uniformly negative. Even *The New York Times* weighed in with an editorial, worrying about the effect Walmart would have on the business:

> There is no chance that Wal-Mart will be buying from small, local organic farmers. Instead, its market influence will speed up the rate at which organic farming comes to resemble conventional farming in scale, mechanization, processing and transportation. For many people, this is the very antithesis of what organic should be. People who think seriously about food have come to realize that "local" is at least as important a word as "organic."

This shifts the battleground of the food wars from a debate between organic and conventional food to a far more esoteric dispute between supporters of the organic movement and those who advocate a far more restrictive standard in favor of locally grown food. According to "locavores," locally grown food is tastier and fresher. But just as important is the fact that buying local is supposedly more environmentally friendly than buying nonlocal organic produce such as strawberries, which may have been shipped thousands of miles from California to New York.

The environmental benefits of local farming are actually highly overstated. Moving food around on ships or by rail is extremely efficient, and once you include the cost of growing, packaging, and even cooking the food, the energy expended on shipping is a relatively minor part of the total energy footprint. In the end, moving locally grown produce around in small bundles, by car or truck to dozens of farmers markets or small retailers, is far more wasteful than putting thousands of tons of bananas on a container ship.

But again, these are the sorts of complications that supporters of local eating are more than eager to sidestep. As *The New York Times* editorial quoted above makes abundantly clear, the standard we are being encouraged to adopt with respect to our food choices has nothing to do with the environmental effects of producing the food or the health effects of eating it. Second, and more crucially, the "local" standard is one that, unlike organic, cannot be met cheaply (such as through economies of scale). Rather, what the editorial rejects about large-scale organic farming is that it resembles conventional production "in scale, mechanization, processing and transportation," which is just an elliptical way of saying that organic has lost its status. The more the organic resembles the conventional, and the more it is available at low cost to millions of everyday consumers, the less it is able to serve as a source of distinction.

More than anything, the loss of distinction is what explains the sudden shift away from the organic to the local. The virtue of localism in all its forms is that it promises to restore the lost status by ratcheting up the stakes: the standard now is one of boutique consumption of goods that are, by definition, more expensive and harder to find than the stuff that any old shopper can find at the supermarket, or the mall. At the extreme end, the seriously committed local shopper acquires everything through social connections – the shame of actually buying stuff on the open market is left for the lesser folks.

A nice illustration of the one-upmanship at work here is the recent fetish for the 100-mile diet, a relatively casual experiment in local(ish) eating by a couple from British Columbia that began in 2004 and that was later turned into a best-selling book. It was quickly adopted all over by adherents of the small-footprint ecological movement, but here's the thing: if it is the virtues of localism you're after, there's absolutely nothing special about 100 miles. Why not, say, 50 miles? And sure enough, there's a website out there for people who want to try a 50-mile diet. But if 50 miles is more authentic than 100, wouldn't zero be most authentic of all? Of course, which is why there is a man from British Columbia named Dan Jason who thinks expanding your graze to 100 miles is for sellouts. But speaking of selling things, for $36 he'll sell you a kit full of seeds so you can grow everything you need in your own backyard or rooftop garden.

This is all part of one of the greatest fads of this young millennium, the trend of turning environmental authenticity-seeking into a competitive anticonsumption publicity stunt. In January 2006, a group of environmentally conscious friends in San Francisco decided to see whether they could collectively go an entire year without buying anything new. They would have to either borrow, barter, or buy second-hand anything they wanted, with the exception of health and safety necessities such as toilet paper and underwear. They called themselves "The Compact," and the movement quickly attracted thousands of like-minded adherents. It also attracted a number of critics, many of whom resented the moralizing tone of the challenge. Others pointed out that by excluding services, especially services like air travel (one member of the Compact took a vacation to Israel), it hardly went far enough toward minimizing the impact of humans on the environment.

Since then, many others have tried to outdo the Compact. Novelist Barbara Kingsolver and her family spent a year eating food produced only on their farm in Appalachia, an adventure detailed in her 2007 book *Animal, Vegetable, Miracle.* Then there is Novella Carpenter's 2009 book *Farm City,* about her adventures raising pigs and rabbits in inner-city Oakland. But for rank absurdity, it is hard to beat *No Impact Man,* a story of a writer named Colin Beavan who convinced his wife and daughter to spend a year in their New York City apartment without producing any environmental impact. No net carbon emissions, no air pollution, no water toxins, no landfill inputs, nothing at all.

The project is full of dumb exercises, most of which are totally disconnected from any actual environmentally sound agenda. For instance, there was the time Beavan accused his wife of cheating by buying a newspaper (newsprint is a carbon sink, after all). More insane was the day he ended up climbing 124 flights of stairs, nine more than the Empire State Building. How many extra calories did he and his family burn by refusing to use the elevator? How much energy was used growing and delivering and cooking the extra food they had to consume as a result? What was the impact of that on the planet?

Beavan doesn't know because ultimately he doesn't care. *No Impact Man* is just the culmination of a form of authenticity seeking whose underlying dynamic is finally revealed as completely absurd. It's like that scene in *Something About Mary,* where Ben Stiller is talking to a stranger he picked up while driving. The stranger tells him that he has a plan to create a new exercise tape called "seven-minute abs," to compete with the then-popular program "eight-minute abs." Stiller looks at the stranger and, quite reasonably, asks him what happens if someone comes along with a program for six-minute abs.

Let us take a moment to remind ourselves of the features and characteristics that constitute the authentic. With respect to the self, it is about being spontaneous, risky, sentimental, and creative. Authentic social relations are built around small, organic communities that are nonhierarchical, noncommercial, and nonexploitative. Finally, an authentic economy is small and low impact, with goods and services sourced locally and provided by family-run businesses, not franchises or other large industrial operations.

It is not difficult to see how every single one of these features or characteristics can serve as the source of authenticity one-upmanship. Being spontaneous, living in a small community, shopping locally – all of these are subject to an indefinite number of divisions and gradations. When it comes to shopping locally, how local is local enough? How risky is risky enough? If we want to live a low-impact, environmentally conscious lifestyle, how far do we need to go? Being an authentic person, or living an authentic life, turns out to be not so different from being a nonconformist: it is a positional good that derives its value from the force of invidious comparison. You can only be a truly authentic person as long as most of the people around you are not. In many ways, the quest for authenticity is just a deeper and more all-encompassing variation on the quest to be cool. Where cool was about nonconformity and the rejection of mass society, authenticity is a root-and-branch reaction against the entire social, economic, and political infrastructure of modernity.

Recognizing that authenticity is a positional good with a built-in self-radicalizing dynamic helps us make sense of a lot of the seemingly bizarre behavior that manifests itself as authenticity-seeking. The fetish for the public display of emotion, which exploded into our popular culture with the death of Princess Diana and whose embers are tended by Oprah Winfrey's cult of self-obsessed sentimentality, can be understood as a form of radically conspicuous authenticity. The pathological concern over the origin and content of our food, accompanied by an almost religious belief in the evils of industrial production and the virtues of local farming, has a similar etiology. The hysteria over global warming that has led to calls for North Americans to give up flying, give up driving, give up meat, give up toilet paper, give up lightbulbs, and give up pro-creating is almost entirely driven by a ratchet of authenticity-seeking that progressively rejects more and more of the comforts and privileges of modern life. Next thing you know, the hyper-rich are sleeping on mud floors, like poverty-stricken Aboriginals in the outback.

More importantly, it is only once we understand that the desire for authenticity is an essentially Veblenian social conceit that we can understand what motivates the distinction between authenticity and "authenticity," or between the genuine form and the fake. It is nothing more than the difference between a type of conspicuous waste where the utter uselessness of the activity remains implicit or unacknowledged, and one where its function as a site of status competition has become cringe-inducingly explicit. That is to say, the dispute that Joshua Glenn flags, between the authentic and the fake-authentic, is just a replay of the ancient hostility between old and new money, between true aristocrats and *arrivistes*, or between people who are cool and those who are trying too hard.

This is why the problem is not that whenever someone invokes the term "authenticity" you know that you've entered the world of the fake authentic. When Levi's promotes its new line of classic jeans as "authentic," or when Dovgan uses the phrase "Dovgan is Authentic Russian Vodka" in its advertising campaigns, at least you know what is going on. In a sense, everything is on the up-and-up and nothing is being hidden. Yes, you are in the world of bullshit, but you know it is bullshit, and they know that you know it is bull-shit. There's a refreshing honesty in the explicitness of the fake authentic, an acknowledgment by everyone involved that there is nothing more at stake than the buying and selling of jeans or vodka.

The implicit authentic is far more pernicious, because its status as a genuine experience or product or service is not in question. We can safely ignore people who go to a restaurant that advertises "authentic Italian cuisine," and we can laugh at friends who rent an "authentic log cabin" on a crowded lake in cottage country. What we need to worry about are the people who go to special invitation-only set-menu dinners hosted by professional Italian chefs or who own cabins on remote, closed-development lakes in Northern Ontario or the Gulf Islands. These are the people who are setting the bar for everyone else; their privilege does not manifest itself as mere privilege, but as the successful discovery of the rare fruit of authenticity. It is precisely because only a few people can partake of this sort of implicit, genuine authenticity that there is a market for the more explicit, fake kind. Just as the phenomenon of keeping up with the Joneses needs to be blamed on the Joneses for starting the competition in the first place, the one-way ratchet of the search for authenticity is the fault of those who set the bar, not those who try to meet it.

FIVE

PERILS OF TRANSPARENCY

Why does Oprah Winfrey keep getting duped by fake memoirists? Herman Rosenblat wrote an autobiographical book about a young man in a concentration camp who was thrown an apple a day by a young woman across the fence; later they met again in Brooklyn and fell in love. Margaret Seltzer wrote about growing up in an L.A. ghetto. Misha Defonseca penned an incredible memoir about a little girl who fled from the Nazis and was – seriously – raised by wolves. All three found themselves celebrated by Oprah Winfrey, either on her show or in her magazine. And all three of them were eventually exposed as grade A fabulists.

But for A+ fabulism, we need to turn to James Frey, author of the memoir *A Million Little Pieces.* In the winter of 2007, Frey appeared on *The Oprah Winfrey Show* to confess that the rumors that had been circulating were true, and that much of his book was a mix of hyperbolic exaggeration and outright invention. Frey was forced to come clean after a report (entitled "A million little lies"), published on the website The Smoking Gun, accused him of larding his supposed nonfiction book with "fabrications, falsehoods, and other fakery" concerning his account of his supposed career as an alcoholic, drug abuser, and criminal. Among the more egregious embellishments: Frey had not been arrested fourteen times and had not spent three months in jail; he had not got on a plane without ID, missing four teeth, "covered with a colorful mixture of spit, snot, urine, vomit and blood." Creepiest of all, it came out that Frey had falsely inserted himself into the narrative of the tragic death of Melissa Saunders, a girl from his high school who was killed when a train hit her car one night.

On Oprah's show, Frey conceded that pretty much everything in The Smoking Gun report was accurate and that, consequently, pretty much everything in his book was a pack of lies. Oprah was incensed. "I feel duped," she told Frey. "But more importantly, I feel you betrayed millions of readers." She had plenty of reason to be mad. After all, Frey had Oprah to thank for having millions of readers to betray in the first place. After Oprah chose A Million Little Pieces as a selection for her television book club in the fall of 2005, it became a publishing monster, selling well over two million copies in a few months. At the height of its popularity, the book sold 176,000 copies in one week.

After the initial revelations on The Smoking Gun, Frey appeared on Larry King Live to defend himself, and at the end of the show, Oprah herself made a surprise call to defend Frey. She claimed that what attracted her to the book was not whether every episode and anecdote was true; it was the book's "underlying message of redemption" that mattered. She didn't stick to that defense for long, and she finally turned on Frey, essentially ordering him to appear on her show to confess his sins and explain his behavior.

The Oprah Winfrey brand is built around a cult of authenticity through therapeutic self-disclosure, of the sort promoted by her frequent guest Dr. Phil. As her own website advertises, Oprah "conveys integrity, evokes trust, and promises pleasure." Further, what made Frey's book so powerful – indeed, why the publisher insisted on marketing it as nonfiction – was that it was the raw and unexpurgated account of one man's descent into a hell of drugs and crime and his ultimate redemption through rehab and therapy. When she chose the book for her club, Oprah said that what made the book so compelling was the knowledge that all this stuff *really happened.*

Writing in *The New York Times*, Richard Siklos observed that what made this case so compelling was that Oprah had lent "her own finely honed authenticity" to Frey only to angrily retract it when he was exposed. The whole affair, he wrote, raised a disturbing question: Does authenticity still matter? Siklos concluded that no, it does not, that we live in a culture that has long since lost its grip on reality. Americans have always been fascinated with replicas, imitations, and other forms of mass-produced pseudo-authenticity, and in the place of the truth we're willing to swallow any reasonable facsimile thereof.

In equating factual truth and authenticity, Siklos steps into a familiar trap. What is odd is that Oprah herself fell into it, given that she is the reigning queen of a form of heavily emotive authenticity that is relatively unconcerned with any hard-to-pin-down notion of "historical accuracy." More important for her purposes is that the story serve as an accurate reflection of the teller's ongoing search for his or her true self, and disclosure matters only insofar as it serves this higher goal. She should have stuck to her original instinct about the book's message being more important than any factual inaccuracies, because for Oprah and her acolytes, history is and ought always to be the slave of the passions. Feelings are what matter, feelings above all.

The point, then, is that what makes the whole Oprah versus James Frey affair so odd is this: the fact that Frey made up his book makes it fictional and historically inaccurate, but it does not thereby make it "inauthentic" in an Oprahian sense. In Oprah's world, authenticity is nothing more than a contemporary version of Rousseau's original idea that one's true inner self is not so much discovered as it is invented, which makes the distinction between fiction and nonfiction essentially irrelevant. In that sense, *A Million Little Pieces* is still a perfect fit for Oprah's Book Club.

After his confession on Oprah, Frey wrote a three-page "note to the reader," which was posted on his book's website and which the publisher has promised to include in all subsequent editions of *A Million Little Pieces*. In it, he claimed that the reason he invented events and embellished so many facts was "in order to serve what I felt was the greater purpose of the book" and to "detail the fight addicts and alcoholics experience in their minds and in their bodies." This is just another way of saying that historical truth is not what matters. Instead, what is important is the overall integrity of the narrative and how it serves the deeper truths of the addict's experience.

As it turns out, this is exactly the attitude endorsed by Oprah herself in her infamous phone call to *Larry King*. As she said at the time, what impressed her about the book was not the small matter of whether it was true, but the underlying and universal message of redemption at the heart of Frey's tale. And she wasn't just free-lancing an excuse; this attitude is entirely consistent with Oprah's overall approach toward literature, which is to use writing as a vehicle for therapeutic self-understanding. In what must have been a calculated and profoundly cynical nod at the narcissism at the core of this literary theory, Frey ended his confessional by saying to Oprah, "What is important is that I learn from this and become a better person."

With that, James Frey was doing nothing more than pledging his undimmed allegiance to the mildly cultish form of authenticity that had led Oprah to endorse his book in the first place. It isn't clear that she grasped the irony, which is that Frey's fabulations are not a symbol of the decline of the importance of authenticity, but just the opposite. The outrage over *A Million Little Pieces* is a symptom of a culture that values authenticity more than just about anything else. But why then were so many people upset that Frey had made up large swaths of his narrative? Again, we find the beginnings of an answer in Rousseau.

Rousseau recognized that the making of an authentic self is a group effort. You cannot just impose any old narrative you want onto the events of your life; the story has to seem plausible and sincere if it is to be accepted by others. Like any other creative work, the public will not accept it as authentic if they suspect that the artist has a hidden agenda or ulterior motives. The problem that James Frey encountered was that many people started to feel that his lies and embellishments were aimed less at doing justice to the realities of addiction than at creating a story that would sell. The lesson: invention in the name of art is authentic; invention in the name of profit is fraud.

Thus, we can situate the Frey affair alongside a broader suite of intellectual sins we can call *vices of authenticity*. These include plagiarism, hypocrisy, and gossip. What makes them vices of authenticity is that their vicious nature consists in a discrepancy or lack of accord between the presentation of a false outer self and the reality of the true inner nature.

Our culture is suffering through what appears to be an interminable – and growing – plague of plagiarists. One of the most famous cases of the past few years is that of Kaavya Viswanathan, the Harvard undergrad who had her first novel (a bit of chicklit fluff called *How Opal Mehta Got Kissed, Got Wild, and Got a Life*) pulled from the shelves in 2006. It transpired that Viswanathan had helped herself to prose from as many as five other books, most egregiously from a couple of books by her competitor in the genre, Megan McCafferty. Around the same time and at the other end of the career path, the CEO of defense contractor Raytheon, William Swanson, suffered the indignity of being censured by the company's board of directors after it turned out that his pamphlet, *Swanson's Unwritten Rules of Management*, had been, well, written by someone else. In 1944.

These are hardly isolated cases. The ranks of writers who have been accused of plagiarism over the past few years include (but are far from limited to) Stephen Ambrose, Ann Coulter, Dan Brown, Doris Kearns Goodwin, Alan Dershowitz, Ian McEwan, and Chris Anderson. The ranks also include the thousands of university students who have discovered that cutting and pasting beats researching and writing any day. Anecdotal evidence has been swirling around the academy for years, and by now it is clear that plagiarism in universities is an out-of-control epidemic.

Any hopes that it might be otherwise were dispelled by the results of a survey jointly conducted by professors Julia Christensen Hughes of the University of Guelph, Ontario, and Donald McCabe of Rutgers, and published in 2006. Hughes and McCabe found that 53 per cent of Canadian students admitted to cheating on written assignments at least once in the past year, with 37 per cent admitting to forms of plagiarism, ranging from cutting and pasting a few lines from books or from the Internet to submitting entire papers that had been written by someone else. Things were not much better at the graduate level, where 24 per cent of students confessed to having plagiarized in the last year. Lest anyone think that Canadian students are especially dishonest, numerous studies have shown that American and U.K. students are just as bad, if not worse.

But what is plagiarism exactly? As Richard Posner writes in his *The Little Book of Plagiarism*, plagiarism is typically understood as a literary theft, where one writer passes off the work of another as his or her own. This definition is misleading on at least two counts, though. First, plagiarism is not restricted to the literary realm, and the history of the arts is replete with tales of musical and artistic plagiarism. Second, calling it "theft" raises all sorts of questions, implying that plagiarism is a criminal matter involving the stealing of property.

This too is inaccurate; there is both more and less to plagiarism than is captured by the term "literary theft." It is instead a form of intellectual dishonesty that has the desire for concealment or deception as its aim. We could say that plagiarism is a form of insincerity, where the culprit tries to display an outer image of his or her intellectual acuity or artistic ability that does not reflect his or her inner capability or sensibility.

What unites these offences is that they all involve some form of misrepresentation of the self. Indeed, in many ways, plagiarism is the flip side of forgery: forgers pass off their own work as that of someone else, while plagiarists pass off the work of others as their own. The reason we get hung up about these things is partly financial, but also moral. We give these moral ideals names like sincerity, uniqueness, integrity, and originality, but the motivating principle is what we have been calling an ethic of authenticity. This intimate connection with the authenticity ideal helps explain our intuitions about what does or does not count as plagiarism. We care about plagiarism when originality is an issue, when who is responsible for creating the work in question matters to us. Plagiarism matters when *authorship* matters. We don't really care about originality when it comes to the authorship of textbooks, in certain areas of law, in some science labs, or sometimes even in artistic collectives such as Rembrandt's studio or Andy Warhol's Factory.

So plagiarism requires the participation of the audience: there has to be an attempt at deception or concealment, as well as an audience that cares if it is deceived. As Posner points out, plagiarism may be a fraud visited upon our expectations, but expectations differ. Not only do people disagree over specific cases, but general standards change over time. It is sometimes claimed that plagiarism emerged out of the Romantic cult of the creative individual, but that isn't quite right. While it is certainly true that our attitudes toward authorship and creativity are shaped (and, perhaps, distorted) by the legacy of Romanticism, the metaphysical essence of plagiarism arose somewhat earlier. Richard Posner reminds us that the term itself comes from an old Latin word for kidnapping or plundering, and that it was most commonly used by the ancient Romans to refer to the theft of slaves. But in the seventeenth century the word *plagiarism* was appropriated into the European vernacular to refer to a form of intellectual or artistic pilfery.

Philosopher André Gombay argues that plagiarism entered the early modern vocabulary as part of a more general interest in the grading or evaluating of persons. Gombay finds the definitive statement of this interest in Descartes' *Meditations*, one of the classic texts of Western philosophy. Gombay is most interested in the third meditation, where Descartes offers his first of two proofs for God's existence. The proof is a variation on the "argument from design," and it runs roughly as follows: Descartes observes that as an imperfect and finite being, there is one idea in his mind that he could not have conceived of through the power of his own intellect, namely, the idea of perfection. So, he says, whoever made me must have put this idea in me, and the only possible maker who could put the idea of perfection in me must himself be perfect, that is, God.

This is a toy problem that philosophy professors use to torment undergraduates, and it never convinces anyone. But while it is hardly convincing as a proof of God, Gombay teases out of it a neat principle at the heart of the argument. What Descartes is asserting is this: *The more accomplished the product, the more accomplished the maker.* As Gombay sees it, this principle underwrites a new understanding of the relationship between the self and creativity. When we see a work of some degree of artifice or ingenuity, we assume that whoever made the work must possess an equivalent amount of creative or intellectual power. In Cartesian terms, we believe that there must be a relationship between the internal, formal powers of the maker and the external, objective reality of the product.

Setting aside its merits as an argument for God, we can see how this principle also underwrites our forensic intuitions when it comes to plagiarism. Anyone who has ever taught in a university and had to grade or evaluate student papers knows that plagiarists give themselves away in a limited number of ways. Sometimes they are too free with the cutting and pasting, and the paper ends up with jarring changes in style, voice, or even font. Another tell-tale sign is that the paper refers to the right texts, but to editions different from the ones that were ordered for the course. But the most obvious sign that the paper is not the student's own work is that it is simply too good. If a student is cruising along at a steady B-minus clip, and he or she suddenly submits a paper of such polish and accomplishment that it could be published, the odds are that it already *was* published . . . by someone else. Why? Because *the more accomplished the product, the more accomplished the maker*, and there is just no way that a typical undergraduate is intellectually accomplished enough to be capable of writing publishable philosophy.

Gombay is not claiming that the idea of plagiarism as a form of intellectual fraud emerged fully formed out of nowhere in the seventeenth century. After all, a hundred years previously, British politician Sir Philip Sydney protested that he was no "Pickepurse of another's wit," while in the first century AD Roman poet Martial accused a fellow poet of plagiarism in something like its modern sense. Yet Gombay is surely right in that the seventeenth century oversaw a transformation in our attitudes toward thinking, making, and the self, a transformation that was reflected by a growing concern over the form of intellectual fraud that we call plagiarism.

What does it take to turn an honest, hardworking young student into a lazy, lying cheat? Some educators suggest that plagiarism is a consequence of the increasingly competitive nature of university. The pressure is on, the stakes are climbing, so there is a greater incentive to cheat. Others suggest that students can't be blamed for merely following the lead of the culture at large. After all, some of the most successful people out there have been caught stealing from others on occasion, and their continued success indicates that they haven't exactly been given a failing grade by society. So there is probably not much to be gained by lecturing students on academic values or telling them "they are only cheating themselves." The fact is, plagiarism is so rampant now that it is close to becoming normalized, where students (and, sadly, even profs) no longer see it as a big deal. If students are increasingly dishonest, perhaps it is because the type of dishonesty that plagiarism involves is increasingly tolerated by society.

But the easiest and most obvious explanation for the apparent epidemic of plagiarism is technology. The mass digitization of texts, combined with keyword-based search engines such as Google, has made helping yourself to the works of others a piece of cake. In the old days, if you wanted to copy something out of an obscure journal or book, eventually you had to type it in by hand, which would take almost as much effort as writing something of your own. The digitization of our culture has transformed writing from a hunt-and-peck business into a cut-and-paste affair. The flip side of this development is that the Internet has also made it far easier to catch plagiarists. Back in the olden days, if a professor suspected a student of cheating, he or she would have to dig through the library by hand, or pass the suspect work around to colleagues. Now, if a passage seems out of place in a student's paper, or if a section in a book seems familiar, you can just type a line or two into Google and – presto – you have your smoking gun.

But that explains only why it is easier to plagiarize, not why so many people choose do it. Is it possible that students and professional writers alike are becoming more dishonest? Despite the evidence, I think the answer is no. It is important to recognize that honesty is not something some people possess and others lack, the way some people have blue eyes and others don't. *Honest* is an adjective that applies to individual actions or statements that are elicited (or not) in a way that is highly sensitive to social context and to expected risks and anticipated rewards.

The point is that no one is 100 per cent honest all the time; everyone is disposed to lie or hide the truth in certain circumstances. Indeed, various studies suggest that people actually lie a lot – telling three or four explicit lies every day, not counting the small lies we tell ourselves. This habitual, everyday fibbing is even the premise for a television show. *Lie to Me* stars Tim Roth as a scientist who can tell when people are lying by interpreting "microexpressions," the almost unnoticeable quick looks that play across a person's face when they aren't telling the truth. The science in the show is genuine, based on the work of Dr. Paul Ekman, who pioneered the field of microexpressions and deception detection.

Like most other forms of dishonesty, plagiarism is a crime of opportunity that even the most upstanding person will engage in as long as a motive is present, the risk is low, and the reward high. A useful comparison is with pornography: are fifteen-year-old boys hornier today than they were, say, fifteen years ago? I doubt it. The level of sexual frustration in teenage boys is probably close to a universal constant. So why do they consume so much more pornography than teens did even five or six years ago? Indeed, we could ask the question of the culture as a whole: have we become more prurient over the past decade?

Almost certainly not. All that has changed is that the risk versus reward calculation in the consumption of pornography has become far more congenial to the average person. Before the Internet, acquiring pornography meant that you had to go into a store, spend money, and then find someplace to consume it. The shame of facing the store owner, the cost of the magazines or videos, and the risk of getting caught by your spouse or children were all too much for most people. But now acquiring pornography is simple, anonymous, free, and doesn't involve shoving magazines between the mattress and the boxspring.

There is something similar at work in the universities with respect to plagiarism. The chief vice of most students (like most humans) is not dishonesty but laziness, and when the innately lazy are placed in a highly competitive environment, it is hardly surprising to see them looking for a route to success shorter than the one that leads through the library. (Exam question: Compare and contrast plagiarism among university students with steroid use by athletes.)

Once again, technology alters the landscape. It used to be that a student who wanted to cheat was faced with a number of serious obstacles. For starters, it was hard to find something suitable to plagiarize. Anything in the library was too good: something copied from a journal or book would stick out like a logician at an existentialism conference. It was possible to buy term papers, usually off impoverished grad students, but aside from the cost, this option had some of the same shame-inducing qualities as a visit to VideoFliXXX, since the student you bought the paper off one semester might be your teaching assistant the next. At the end of the day, an all-nighter was often the best, if not only, bet.

None of these obstacles remain. The Internet is chockablock with extremely mediocre academic writing that is perfect for cutting and pasting. A student looking for something a little more polished has access to a number of online paper mills that will custom-produce term papers quickly, cheaply, and anonymously. Why stay up all night when you can head for the pub and have the paper in your inbox at dawn? That is why there is probably not much to be gained by lecturing students on academic values.

That takes care of the students, but what about professional authors? Plagiarism among writers is a bit harder to understand. True, they have many of the same problems as students – laziness, competition, deadlines – but the stakes are way higher. In academia, hardly anyone actually reads most academic papers, so the risk and reward are relatively independent and the reward for a successful plagiarism may seem worth the small risk of getting caught. But in the literary field, the risk rises in lockstep with the reward. For Kaavya Viswanathan, the probability that her plagiarisms would be found out rose to almost certainty the minute it was announced that she had received half a million dollars' advance and that the film rights had been sold to DreamWorks. The general point is that the literary plagiarists' desire to achieve success without effort will be self-defeating, since the more successful they are, the more likely it is they will be caught.

But maybe what motivates the literary plagiarist is something more than the eternal appeal of the easy road to fame and fortune. There is a scene in the movie *The Squid and the Whale* where Walt (the older son of the couple played by Laura Linney and Jeff Daniels) sings a supposedly original song at the school talent show, which he wins. It is an excruciating scene to watch, since the song he plays is the classic rock staple "Hey You" from the Pink Floyd album *The Wall*. The audience knows that even though the movie is set right around the time of the album's release, it is only a matter of time before someone recognizes the tune and rats Walt out.

But remember what Walt says after he's caught? He admits to stealing the song but not because he needed the prize money, or because he was trying to impress a girl or his parents or his classmates. In fact, Walt does not even accept that what he did was wrong, since he felt that it was the sort of song that he *could* have written. That is, the fact that Pink Floyd wrote the song first is irrelevant, since it is an authentic outward expression of Walt's deepest feelings and creative impulses.

The whole episode is hilarious (to the extent that extreme vicarious discomfort is funny), but it also gets at one of the unfortunate realities of creative life. Every writer (and equally, every musician, artist, and even scientist) runs into a situation where she comes across a work that strikes her as so obviously right, and so perfectly constructed, that she feels that she would have done it exactly that way if only she'd thought of it first. On these occasions, passing the work off as your own does not feel like plagiarism; it feels more like the appropriation of something that is a part of your true self.

This is one aspect of the quasi-oedipal phenomenon that literary critic Harold Bloom calls "the anxiety of influence." It is the concern that every artist has at one time or another, that they cannot escape their influences, that they are late to the party, that everything worth saying has been said. So we are confronted with the following possibility: the reason that plagiarism is on the rise, even among professional writers, is not because people are dishonest or because we care less about the morality of misrepresentation but, paradoxically, because we care about it too deeply. Because of our commitment to authenticity, we tend to look down on ideas that are borrowed or derivative. We fight over credit for things, partly because there are potential financial or status rewards, but also because we celebrate creativity and we believe there is something profoundly unjust about people receiving credit for books they didn't write or for inventions they didn't conceive.

The problem is that the more we demand originality, the harder it is to actually be original. This gives us a strong incentive to lie or to conceal the origins of our ideas. How can someone legitimately claim to be a visionary CEO unless he or she has some secret (unwritten!) rules that no one else has? How can a young Harvard student present herself as the voice of her generation if she has to steal ideas from women decades older? The requirements of authenticity have led us to a point where the demands of originality are quite literally impossible. We are all, for better and for worse, creatures of the culture in which we swim, influenced in myriad ways we cannot even begin to sort out.

But over the past decade or so, the culture has undergone a profound shift, as new technologies of networked digital communication have reworked our intuitions about originality, creativity, plagiarism, and authenticity. Intellectual property law is all of a sudden hip, happening, and sexy. Want to know what the kids are all worked up about? The answer is copyright.

As I was writing this chapter, actor Christian Bale was in the news after a recording of an obscenity-laced hissy fit he threw on the *Terminator: Salvation* set was posted on the Internet. Bale went postal on the film's director of photography, Shane Hurlbut, after Hurlbut accidentally entered the actor's sight lines during filming. Around the same time, a short YouTube video was making the rounds that showed a seven-year-old kid named David sitting in the back seat of the family car as his father drove him home from the dentist. David is clearly drugged to the gills, and he keeps asking his father questions like "Is this real life?" and "Why is this happening?" He also tells his father he has four fingers, and at one point he starts screaming for no reason and then quickly falls back in his seat, head lolling.

Aside from the fact that both of these items are, in their own way, both very funny and somewhat creepy, they were also the subject of a huge number of mashups. On blogs, video-sharing sites, and other web portals, users posted versions of Bale's rant mixed with the Huey Lewis song "Hip to Be Square" (a reference to an earlier Bale film, *American Psycho*), mixed the rant with hip-hop beats, and re-enacted the rant with puppets and action figures. One amateur actor even re-enacted the whole thing by directing his rage at a pile of unglazed donuts. As for David, the video of his trippy little ride was quickly set to various 1960s acid-rock tunes, while in another video the whole episode was re-enacted by someone in a Darth Vader costume. And of course, the most popular mashup of all was the entirely predictable video "Christian Bale Takes David to the Dentist."

Thanks to the Internet and cheap digital manipulation software, the old, rigid distinctions between creators and consumers, between mass society and local markets, and thus between commercial and noncommercial culture, have completely dissolved.

High-bandwidth connections and peer-to-peer file-sharing net-works have turned every dorm room into a broadcast studio, while digital technologies have democratized artistic production. Using the omnipresent sea of symbols, images, sounds, and texts as source material, millions of people are taking the rip/mix/burn mantra to heart and laying claim to their cultural inheritance.

In the process, through widespread engagement with social net-working sites and other related forms of communication, they are transforming our understanding of the nature of privacy and per-sonal information while engaging in what is beginning to look like a massive sociological experiment in voluntary disclosure. At the same time, this cultural shift is challenging our assumptions about the authentic self, in particular the familiar Rousseauian fantasy about perfect transparency leading to more intimate and egalitar-ian personal relationships.

It took a while for corporations and governments to twig to what was going on. When they finally clued in around the mid-1990s, the effect was tectonic, causing an Andes of angst in the boardrooms of copyright-rich corporations. The very features of the Internet that made it so appealing to business – the ability to store, alter, and deliver content anywhere on Earth quickly and at no cost – also appeared to make it impossible to enforce intellec-tual property rights.

These are the roiling headwaters of what came to be known as the Copyright Wars, a fast-evolving conflict between copyright holders (some artists, musicians, and writers, but mostly big copyright-rich companies such as Disney and EMI) and content users, who were no longer satisfied playing the role of the passive consumer. They increasingly felt entitled to remix/reuse/recycle elements of the culture as they saw fit, to share their "creations" with friends or on the Internet, and profit from them if possible.

One example (among tens of thousands) that has achieved a certain degree of notoriety was the famous "Hope" picture, created by artist Shepard Fairey during the Barack Obama presidential campaign. It quickly became the defining image of Obama's candidacy and was featured on everything from posters to T-shirts to mugs to buttons. The image was based on a photograph of Obama taken by freelance photographer Mannie Garcia, who worked for Associated Press. After the election (and after reproductions of the image started to generate some revenue), AP laid claim to the work, stating that Fairey's art was an infringement on its copyright.

As the case was reported on blogs and in the newspapers, Fairey was usually accused of plagiarism. But while copyright infringement and plagiarism often overlap, they are not the same thing. Copyright gives its holder the limited right to control the copying of whatever work the right covers, but despite the fact that it falls under the aegis of what is known as intellectual property law, copyright does not confer a property right upon the holder. What we call a copyright is actually a bundle of privileges that give the holder the exclusive right to make copies, create derivative works such as a translation or the novelization of a screenplay, perform it publicly, and transfer or license any of these rights to a third person, such as a publisher or distributor. The publisher of this book currently holds the copyright, which means that no one, not even me, is allowed to make a copy of it without the publisher's permission. When he died, Michael Jackson was co-owner of the publishing copyright to most of the Beatles catalogue. He owned it outright at one point, which was why he was able to license the song "Revolution" to Nike for a 1997 ad campaign.

In the United States, copyright generally lasts for the life of the author plus seventy-five years, after which the work enters the public domain and can be freely copied by anyone. (In Canada, it's fifty years.) Also, anyone can use any work for the purpose of private study, research, criticism, or news reporting, as well as for purposes of satire or parody. This "fair use" provision is what permits academics to cite other works in their articles, and it is what enables reviewers to write critical reviews of books in the newspaper. Without a fair use exemption, copyright holders would be able to effectively shut down public discussion or debate about their writing.

Plagiarism and copyright infringement are often confused, largely because plagiarists frequently help themselves to copyrighted works. Nevertheless, not all plagiarism makes use of copyrighted material, and not all copyright infringement involves plagiarism. If I pass off material from the public domain as my own work, I'm a plagiarist, but I have not violated any copyright. On the other hand, if in writing a book I help myself to thousands of words from another work, the fact that I put quotation marks around the borrowed passages will save me from the charge of plagiarism but will put me beyond the limits of what constitutes fair use. To put it simply, copyright infringement is a legal question, while plagiarism is a moral offense.

But underlying these issues is a deeper question: to whom, ultimately, does the culture belong? Whenever a copyright holder claims infringement, or a writer or photographer protests that they've been plagiarized, they are laying claim to ownership or authorship – they are insisting that they have a privileged connection to the work in question that gives them a right to control how words or images or symbols are circulated. In both cases, the claim acts as a brake, stifling creativity, smothering innovation, and squandering the democratic potential of the new technologies. As law professor and copyright reform activist Lawrence Lessig has written, "What's at stake is our freedom – freedom to create, freedom to build, and, ultimately, freedom to imagine."

Lessig frames the discussion in terms of a broader cultural shift, driven by technological evolution, from what he calls a Read Only (RO) culture to a Read/Write (RW) culture. An RO culture is one that most people over the age of thirty are most familiar with. It is the culture dominated by what is commonly referred to as the "mass media" or, in more derisive terminology, the "mainstream media," or MSM. This is the culture of mass-market newspapers and magazines, broadcast television networks, and large corporate music labels and film studios. An RO culture is characterized by a sharp distinction between producers and consumers and is usually propagated by a unidirectional one-to-many mode of transmission. While these aspects may be seen as drawbacks, the old RO culture has a number of important virtues. In particular, it fosters professionalism and authoritativeness and has developed effective mechanisms of quality control and accountability.

But the MSM is in trouble. The newspaper sector in particular is being decimated by competition from blogs, news aggregators, gossip sites, and classified ad services such as Craigslist and eBay. But it isn't just print media that are hurting; music studios are losing out to file sharing services, iTunes, and online radio, while television is losing viewers to YouTube and Hulu. The entire business model of the old mass media has come under incredible strain, and in the case of newspapers it has completely shattered.

These businesses are going to be difficult, if not impossible, to rebuild. It isn't a simple matter of translating the old business model to new media by finding a way to "monetize" the readers and viewers who have migrated to the Web because the whole texture of the culture has changed. We have shifted to an RW culture, which has a completely different set of values and imperatives. In an RW culture, the old distinction between producers and consumers disappears, as people feel free to edit, remix, and recreate at will. In contrast with an RO culture, an RW culture values amateurism, sharing, and collaboration. It is more like a dialogue or conversation than a lecture, and quality control is maintained by an open-source editing and multiple drafts approach.

While this may seem like something completely new, Lessig stresses that the recent move from RO to RW is actually a return to the norms that dominated the culture up to the end of the nine-teenth century. The first shift, from an earlier RW culture to the RO establishment that ruled for most of the twentieth century, was itself sparked by a technological revolution that for the first time allowed sounds and images to be preserved, copied, and sold or broadcast to the masses. Player pianos, phonographs, radios, film reels – all of these undermined the mongrel and "open-source" folk traditions that flourished in the great melting pot of the United States. The new mass media took the culture out of the hands of the people and gave control to a professional elite or cultural monarchy. And so when we look back over the past couple of centuries, the heyday of the RO culture after the Second World War looks like a relatively brief inter-regnum, an aberration that was made possible by a distinct and tem-porary level of technological development of analog mass media. The digital revolution is taking us back to a more democratic culture built around creative communities of sharing and collaboration.

Not everyone is happy to see the dinosaurs of the MSM lumber-ing toward extinction. One of the best attacks on the values of RW culture is Andrew Keen's book *The Cult of the Amateur*, a splendidly written rant against Web 2.0 applications and services, from blog-ging and YouTube to file sharing and Wikipedia. As Keen sees it, the amateur and collaborative RW culture of which Lessig is so fond is actually a cesspool of ignorance, egoism, bad taste, and mob rule. He laments the loss of the professionalism, integrity, and author-ity of the old RO institutions and weeps for a world where the "monkeys" have taken over:

> Say good-bye to today's experts and cultural gatekeepers – our reporters, news anchors, editors, music companies, and Hollywood movie studios. In today's cult of the amateur, the monkeys are running the show. With their infinite typewriters they are author-ing the future. And we may not like how it reads.

Keen is certain that he does not like it. As far as he is concerned, there is nothing "democratic" about the new media. If anything, they work to undermine democracy by helping spread propaganda and lies through fake videos, guerilla and viral marketing campaigns, and the widespread use of "sock puppets," which are fake online identities that are used to promote a cause or a product under the guise of independent support or endorsement. There are too many places to hide online, Keen argues, and what we are left with is not the democratic wisdom of crowds but an idiocratic mob rule, a cynical and infantilized political culture and a degenerate civil society.

The sharpest version of the argument that the Internet is bad for democracy comes from Cass Sunstein, a law professor at the University of Chicago. In recent years, Sunstein has been fussing about the rise of what he calls the "Daily Me," the way the Internet permits highly personalized and customized information feeds that guarantee that you will be confronted only with topics that interest you; they screen out those that may bore, anger, or annoy you. As Sunstein sees it, the Daily Me harms democracy because of a phenomenon called group polarization: when like-minded people find themselves speaking only with one another, they get into a cycle of ideological reinforcement where they end up endorsing positions far more extreme than the ones they started with.

Group polarization is a real phenomenon, and it turns up in all sorts of places. It helps explain why, for example, humanities departments are so left wing, why fraternities are so sexist, and why journalists drink so much. For the most part, group polarization isn't a problem (for democracy anyway), since over the course of our day we routinely come into contact with so many people from so many different groups that the tendency toward polarization in one is at least somewhat tempered by our encounters with others.

But Sunstein is worried that group polarization on the Internet will prove far more pernicious. Why? Because the blogosphere functions as a series of echo chambers, where every viewpoint is repeated and amplified to a hysterical pitch. As our politics moves online, he thinks we'll end up with a public sphere that is partisan and extreme. As an example, he points out that 80 per cent of readers of the left-wing blog Daily Kos are Democrats, while fewer than one per cent are Republicans. The result, he claims, "will be serious obstacles not merely to civility but also to mutual understanding."

The Keen-Sunstein reaction to new media is ingeniously topsy-turvy. For decades, progressive critics have complained about the antidemocratic influence of the mass media of the RO culture because of the way newspapers and television networks have presented a selective and highly biased picture of the world, promoting pseudo-arguments that give the illusion of debate while preserving the status quo. For a reminder of how much things have changed, remember that the villain in *Manufacturing Consent*, the 1992 documentary about Noam Chomsky's crusade against media bias was – wait for it – *The New York Times*. And now that the Internet is poised to cast these devourers of black ink and dead trees into the pit of extinction, everyone is up in arms about the consequences for democracy. As a sign of how much the political left has lost its mind over new media, consider that in early 2009, the U.S. Congress was seriously considering support measures for the print industry that included direct federal subsidies, regulatory reform to make newspapers nonprofit enterprises, and, most outrageously, a suspension of antitrust laws so that papers could establish an online pricing cartel.

The error here involves a misunderstanding of how the marketplace of ideas works. There is no reason at all to be concerned that 80 per cent of Daily Kos readers are Democrats, any more than we should worry that probably 80 per cent of the visitors to McDonald's like hamburgers. Given what each of these outlets is selling, it would be bizarre if it were otherwise. If four-fifths of Democrats read only the Daily Kos, that would be worrisome, just as it would be a matter of serious public health concern if 80 per cent of the population ate only at McDonald's. But there is absolutely no evidence that this is the case.

In early 2008, the Project for Excellence in Journalism (PEJ), a think tank sponsored by the Pew Charitable Trusts, released its fifth annual report on the state of the news media. For the most part, its analysis of the newspaper business confirmed the trends of declining circulation, revenues, and staff. But with respect to public attitudes, the PEJ found that most readers see their news-paper as increasingly biased, and that 68 per cent say they prefer to get their news from sources that don't have a point of view. The PEJ also found a substantial disconnect between the issues and events that dominate the news hole (e.g., the Iraq surge, the mas-sacre at Virginia Tech) and what the public wants to see covered – issues such as education, transportation, religion, and health. What this suggests is that readers go online in search of *less* bias, not the comforting self-absorption of the Daily Me. Nothing about how people consume media online suggests they are looking for con-firmation of pre-existing biases. In fact, we have every reason to believe that as people migrate online, it will be to seek out sources of information that they perceive to be unbiased, and which give them news they can't get anywhere else. The mainstream media may be dying, but in the end our democracy will probably be healthier for it.

There is a deeper worry about RW culture, though, which goes beyond its immediate effect on the economics of the MSM, or even on democracy. The worry is that Sunstein's Daily Me or Keen's cult of the amateur are, in fact, offshoots of the Oprah-fied authentic-ity, with its self-absorbed emphasis on emotional truthiness. This is a world where truth has ceased to have any connection to reality, where each person's "truth" becomes as valid as any other person's, and where the only thing that matters is the isolated, self-contained ego – You, and your Feelings.

As Andrew Keen sees it, the moral center of the Web 2.0 revolution is Rousseau's conception of the state of nature as filtered through Oprah's cult of sentimentality, heralding "the triumph of innocence over experience, of romanticism over the commonsense wisdom of the Enlightenment." Nothing, he thinks, captures this triumph of the self better than *Time* magazine's 2006 declaration that the Person of the Year was . . . You, the millions of independent users who have used the Internet to seize the reins of the global media and frame a new digital democracy. The inevitable result of this worship of the self is self-worship, as broadcasting turns to ego-casting and each person becomes a solitary producer and their own solitary audience.

There is no question that the rise of the digital technologies has motivated all sorts of fantasies about the liberation of the individual from the chains of mass society, beginning with the post-1960s cyber-libertarianism that flourished in the early, freewheeling days of the Internet. But critics such as Sunstein and Keen get it wrong when they suggest that the ultimate goal of the new RW culture is isolation, solitude, and narcissism. The fact is, the self is not king, *connectivity* is, and the real social revolution – and the real commitment to authenticity – isn't found in the self-absorption of the Daily Me, but in the perpetual and utterly transparent We of social networking.

In 1785, philosopher and social reformer Jeremy Bentham introduced a design for a new type of prison he called the Panopticon, which consisted of a donut-shaped building with an observation tower in the center. The main building would be divided into cells, with each cell extending the entire thickness of the building, with windows at the front and back. The prisoners would then be backlit, while the windows of the prison tower would be shielded by Venetian blinds. In such a system, the prisoners would be on constant display, while the guards in the tower would remain unseen. The prisoners would always be potentially subject to observation, but they would never be able to tell when they actually were being watched. The whole setup would be like Santa Claus for criminals: they would have no choice but to always act as if they were being watched, modifying their behavior accordingly.

As Bentham saw it, the Panopticon was a clear improvement on existing prisons. To begin with, it would be a lot cheaper, since it would require fewer staff. In effect, the architecture becomes the jailer, with the prisoners watching themselves. But more importantly, Bentham hoped that the Panopticon would lead to moral reform, as criminals, low-lifes, deadbeats, perverts, and other sociopaths would find they had no place to hide. Thanks to the Panopticon, he wrote, "Morals reformed – health preserved – industry invigorated – instruction diffused – public burthens lightened – Economy seated, as it were, upon a rock – the Gordian knot of the poor-law not cut, but untied – all by a simple idea in Architecture."

It has been clear for a while now that we are rapidly moving toward a sort of society-wide "information Panopticon," where technological tools of monitoring and surveillance substitute for the architecture of the prison. As far back as 1999, Sun Microsystems CEO Scott McNealy got testy at a new product launch over questions from a reporter about privacy safeguards, and he promptly announced that in the online world, privacy concerns were a nonissue. "You have zero privacy anyway," he said, adding that people needed to "get over it." McNealy's offhand remark immediately became a sort of antirallying cry for consumer advocates, privacy watchdogs, and media critics who were increasingly concerned about the amount of personal information that was being harvested online and shared among corporations, governments, and policing and security agencies.

The past decade has pretty much vindicated McNealy's take on the situation, though not in the way most people expected. It is

certainly true that we are now thoroughly immersed in a surveillance society and that some of this surveillance is crudely Big
Brotherish, as in the increasing prevalence of closed-circuit security cameras in urban areas. Meanwhile, when Google introduced
Latitude, a mapping application that would provide users with a
map of their friends' locations anytime and anywhere, there were
some perfunctory protestations. A few newspapers led their stories
about Latitude with "watch out, errant husbands" anecdotes; others
raised concerns about the workplace and how owners might use
the technology to implement Stasi-inspired surveillance regimes
on their unsuspecting employees.

Similar privacy concerns have been raised about another
Google service called Street View, a mapping tool that offers 360-
degree panoramic views of urban areas. But for the most part,
Latitude and Street View have been met with a quick shiver of
wary interest followed by one big shrug of collective indifference,
a sort of acceptance of the inevitable erosion of another few feet
in the cliff-face of the headlands of privacy. Google, as we are starting to realize, is not a search engine, it is a surveillance engine. In
fact, the entire Internet itself is turning into one big surveillance
engine. But it isn't Big Brother watching us. It is not even, as some
culture jamming types like to say, a new culture of "sousveillance,"
with millions of "little brothers" watching back and keeping the
government and corporations at bay. Far more pernicious, because
it is more subtle, is the way that we have been easily adopting
devices and sliding into habits that allow us to effectively keep
tabs on one another.

It is by now a cliché to note the ridiculously casual approach
kids these days have toward privacy, especially on their personal
online journals and social networking tools such as Facebook,
Flickr, and whatever successor application is all the rage by the
time this book is published. It is routine to hear about relationships
that have broken apart, friendships that have dissolved, and job

opportunities that have been lost because thoughts or pictures or videos that not long ago would have been considered deeply personal have been offhandedly posted online for mass consumption. Tens of millions of people think nothing of disclosing the most intimate, banal, and embarrassing facts of their daily lives, and what it resembles more than anything is Rousseau's old dream of a self that is utterly transparent to the world. We can think of the current fad for total online disclosure as a massive *experiment in authenticity*, where nothing is hidden and we lay it out on the table, warts and all. "Take me as I am," we seem to be saying, "for I can be no other."

This fits with popular usage, where it is common to use the term *authentic* to describe someone who speaks his or her mind and who refuses to self-censor in deference to social mores or standards of good taste. We admire the apparent spontaneity of someone who tells us how they really feel, instead of hiding behind a mask of appearances and feigned sincerity. This is the source, for example, of John McCain's longstanding reputation as one of America's most "authentic" politicians. When he ran for president the first time in 2000, and again in 2008, his campaign bus was called the Straight Talk Express. More than almost any other politician, McCain prefers an open-access media strategy when on the stump, and he frequently shocks reporters with his candor and lack of guile.

So what is it like, living in a world where John McCain is a role model, where complete candor and disclosure is seen as an overriding moral virtue? In a sense, what we're seeing is a great sociological experiment in the slow but steadily corrosive effects of information technology on the private sphere. For all their benefits, digital communications technologies work as a sort of social acid, eating away at the boundaries between public and private and eliminating the established norms of discretion, courtesy, and common sense. We aren't living in a Benthamite prison, a totalitarian state of panoptical surveillance. No, where we've arrived is a place much more worrisome. Our entire lives now resemble the teenaged prison called high school, where the predominant form of discourse is something known as gossip.

People have long had bad things to say about gossip, and for people forty-ish and over, it is still considered a major social solecism. Gossip is often defined as malicious or harmful talk about the private affairs of others, and gossiping is usual derided as the idle chat of idle women, exchanging petty tidbits about their friends and neighbors as they fold laundry or drink tea or do the shopping. This sexist spin is not entirely unjustified: gossip is a form of manipulation usually exercised by the weak against the strong, and for centuries it provided women – especially those at court or near other centers of official power – with alternative and subversive forms of power against the more conventional authority of men. Women have access to deep secrets about men, which gives them a real, if subtle, bit of leverage over even the most powerful ruler and fearsome tyrant.

But there is another defense of gossip that flags its ability to reveal to us the elusive, quirky, and shadowy truths about human nature. As philosopher Ronald de Sousa writes, "Gossip is inherently democratic, concerned with private life rather than public issues, 'idle,' in the sense that it is not instrumental or goal oriented. Yet it can serve to expand our consciousness of what life is about in ways that are effectively inaccessible to other modes of inquiry."

This is the counterpart to the argument that gossip subverts traditional hierarchies: by cracking open the private sphere to general public scrutiny, gossip can be an instrument of egalitarianism and social leveling. It reminds us that everyone is flawed in some way, that we all have habits, desires, beliefs, and character traits that are unpleasant or perverted, and that we are all insecure or cowardly or vain. That we are all, in a very base and common way, human. Because of the way it can cut the legs from the public stilts we build so that we might strut around a bit higher than our fellow citizens, we may want to call gossip a *virtue of authenticity*.

As an example of this virtue in action, de Sousa cites the groundbreaking Kinsey reports on sexuality, which he describes as a kind of systematic gossip. What was so astonishing about the reports was the way they revealed to American society the way many behaviors – constant masturbation, sadomasochistic fantasies, extramarital sex, homosexuality – that everyone thought were abnormal and perverted were actually extremely widespread (indeed, "normal"). De Sousa expands on this, arguing that if *all* relevant truths were made public, it would become much more difficult to harm someone by disseminating facts about their supposedly "private" affairs. In such a utopia of authenticity there would be no more shame, no more hypocrisy.

Indeed, one of the more disturbing bits of social fallout from the great Internet explosion of the last fifteen years is the realization of just how much sexual deviance is out there. The point is not to justify this sort of behavior, or to argue that just because it is more widespread than we thought, it shouldn't be illegal. To know all may not be to forgive all. But at the very least, a wider dissemination of this kind of knowledge would mean we could stop poisoning our interpersonal lives through deception and betrayal.

In many ways we are coming close to de Sousa's utopia, as the trend toward full disclosure online, combined with the ease with which information can be linked to or passed around, has turned the Web into a giant gossip sheet where, as Scott McNealy told us a decade ago, there is no privacy. And we are far from taking his advice to "get over it." At the very least, we don't seem to have reached the point where there are no longer any secrets and lies. What we are seeing, instead, is a whole lot of anguish and embarrassment, not to mention lost jobs, broken marriages, lost friendships, and lawsuits. It could be that we are simply in a transitional period, with one foot in the old world and one foot in the new, caught in a psychological tug-of-war between two conflicting sets of values. Perhaps the millennial generation, the Gen Y cohort that has never known a time before social networking, will lead the way into a weird sort of future where everything there is to know about everyone is freely available and instantly accessible, and no one will find this openness remotely odd or unpleasant.

Or perhaps there is a flaw in the ideal of authenticity as complete disclosure. To see what it might be, let's think back to why transparency was ever seen as a virtue in the first place. Remember that for Rousseau, the aspiration was to bring your deepest thoughts, feelings, and aspirations into line with how others perceive you. That is, the authentic project is to bring the outer appearance and inner sentiment into alignment, to become who you feel yourself to be, and to have others recognize you as that person. But this project was never about merely telling facts about yourself. More than that, it was seen as a moral achievement, the result of a long, arduous, and artistic process of self-creation.

That is why the Facebook-style habit of promiscuous disclosure very much misses the point, because what makes transparency and openness valuable is that there is a voluntary and somewhat discretionary aspect to it. To put the point bluntly: what matters about secrets is not whether you keep them, but who you choose to tell them to. Telling secrets or embarrassing facts or hidden truths about your past is a way of building trust and fostering intimacy – that's why one of the first things people do (or at least, used to do) when they start dating is show each other their photo albums. As we learned from *Seinfeld*, when a friend tells you a secret and asks you to "put it in the vault," it means you can't tell anyone. Except, that is, your partner or spouse, whom *you have to tell*. It's one of the basic rules of being in a committed relationship.

Honesty, trust, intimacy, discretion: these are adult virtues, and there's a reason why the TV show *Gossip Girl* is set in high school. But they are also the public virtues of a liberal democracy, founded on the promise that there is a distinct, if constantly evolving, line between matters that are in the public interest and those that are, and ought to remain, private. The problem with easy, instant, and ubiquitous surveillance – that is, the problem with the rise of the Gossip Society – is that when the default assumption is that everything ought to be shared, someone who tries to defend privacy inevitably starts to sound like they are trying to defend the right to goof off at work or sneak around on their spouse.

So to defend privacy, we need to understand why it is valuable for its own sake. In particular, we need to appreciate the way that a robust respect for privacy underpins most of the freedoms we take for granted in a liberal society. As Lawrence Lessig has been arguing for years, the United States and Canada actually have a system of internal passport control that is potentially more draconian as anything ever implemented under Soviet Russia or in Apartheid South Africa. It is called a driver's license – a document that tells the state who you are, where you're from, your criminal background, and other similar details. The simple driver's license gives the state the knowledge and the mechanism to impose serious internal controls on the population, if it so chose. It could simply pass a law entitling all public agents to require you to produce your license on demand.

But the sheer expense of such control would be tremendous. Forcing people to register with the authorities when they move around, or putting checkpoints on every corner, is just too costly, both politically and financially. And as Lessig likes to point out, this costliness, this *enforcement friction*, is a source of great political liberty and psychological freedom, which emerges as a happy byproduct of the fact that enforcing the letter of the law is often just too much of a hassle.

Which raises the question of why so many people obey the law, or recognize moral constraints, even when they know there is no chance of getting caught. Philosopher Immanuel Kant drew a distinction between acting from duty and acting in accordance with duty, which is just the distinction between doing the right thing because it is right and doing the right thing because you might get caught. For Kant, only actions done from duty count as moral. But more importantly, he saw that the way obeying the rules (or what philosophers call "deontic constraint") can be a type of freedom. The freedom Kant is getting at here is not the freedom to cheat on your spouse or your employer, it is the freedom of an autonomous will, which chooses to follow the rules. Ultimately, privacy is a value for a culture of people who believe in autonomy, judgment, and personal responsibility and who follow the law or do good out of a sense of duty or moral judgment, not out of a fear of being caught or out of knowledge that they are being watched.

As in the law so in the personal realm. Sometime soon, constant surveillance will become almost costless, and each company will be able to track listless workers, and every spouse will be able to track his or her wayward partner. It could very well be the end of laziness and the end of infidelity, and there may be some public benefit in that. But there's no dignity in doing your job or remaining faithful simply because you have no option. Thus the end of privacy will also mark the end of our cultural adulthood. Surveillance is for criminals and pointless gossip is for children, and it may turn out that "getting over" privacy will mean getting over the dream of liberal democracy.

—

SIX

VOTE FOR ME, I'M AUTHENTIC

CANADIANS ARE STRONG BELIEVERS IN DEMOCRACY. IN KEEPING with the lengthy constitutional heritage that we inherited from the British, we are committed to the principles of individual rights, the rule of law, respect for minorities, and representative government by a Parliament elected through universal adult suffrage. We even spend a great deal of time, effort, and money promoting democracy abroad, and over the past few years Canada stood fast and strong in support of the popular revolutions that developed in Eastern Europe, especially the Orange Revolution in Ukraine, where Canadians played a key role as independent observers.

Yes, Canadians are deeply committed to democracy – as long as it is being practiced someplace else. Here at home, we have developed such a firm contempt for the democratic process that in our national elections, virtually every issue of substance, from our role in the war in Afghanistan to the perennial question of Quebec's place in the federation, is overshadowed by the media's interest in people who have little interest in the election and who show no inclination to vote. Instead of being treated as the bad citizens they are, these "alienated voters" are routinely heralded as canaries in the coal mine, principled abstainers from a sham democracy.

This is not just a Canadian problem. With a very few exceptions, voter turnout is in steady decline in all Western democracies. The phenomenon is commonly referred to as "voter apathy," although it seems to be rooted in a much more active dislike of politicians and the political system. In all the stories you see about people who do not vote, it is rare to see anyone admit to being simply uninterested in politics. Virtually every person profiled has something to say about the issues, the parties, the leaders, and the system. It is just that none of it is positive.

The complaints are built around a common theme, which is that the democratic system provides only the illusion of choice. "There is no real difference between the political parties" is one typical complaint; another is that "none of the parties speaks to me or reaches out to me or represents my views." A variation is that politicians are liars and phonies who are not open to new ideas, never do what they say they will do, which is itself read as a natural consequence of the fact that politics is shallow and image obsessed, driven by spin doctors and pollsters who see their craft as just another branch of marketing.

Consider the case of former, and now disgraced, U.S. senator John Edwards. In the spring of 2007, when Edwards was running for the Democratic Party nomination for president, he was obliged to return $800 to his campaign to cover the cost of two haircuts from a Beverly Hills stylist. For most people, the revelation that they routinely spend $400 on a haircut would be a bit embarrassing, except that Edwards is notorious for being inordinately proud of his hair. So much so that when he ran for vice-president in 2004 as Al Gore's running mate, Republicans took to calling him "the Breck girl" (after the famous ad campaign from the 1970s for Breck shampoo that featured models with luxuriantly feathered hair).

Canadians like to think that they hold the moral high ground over their American cousins, in politics as in so many other aspects of life. But even though we had a good laugh at Edwards for epitomizing the image-obsessed shallowness of American political culture, around the same time that Edwards was taking heat, the

Canadian press was starting to ask questions about Michelle Muntean, an image consultant and psychic who served as a full-time personal stylist for Stephen Harper, the Conservative prime minister. The discovery that he had an image consultant only confirmed what critics had been saying about Harper since he took power in 2006, namely, that he was enamored with the United States and its republican institutions, especially the pomp and deference accorded the presidency.

This sort of stuff also tends to confirm what just about everyone believes about politics these days: our leaders are little more than talking heads, blow-dried actors whose performances are stage-managed by a phalanx of speechwriters, spin doctors, marketing gurus, pollsters, and image consultants. While these people may have once been content to stay behind the scenes, strategists and consultants are now household names. For example, everyone knows that Karl Rove is the evil genius responsible for George W. Bush's two victories, while the political consultant James Carville is a public figure who has written a stack of books about how he and his team engineered Bill Clinton's back-to-back presidencies.

The stories of image-obsessed pols pile up. Al Gore notoriously asked Naomi Wolf, author of *The Beauty Myth*, to give him style advice during his 2004 campaign (she suggested he wear earth tones), and Bill Clinton kept Air Force One sitting on the runway at LAX while he got a $200 trim from Christophe, a Belgian hairdresser-to-the-stars, but it is important to keep in mind that none of this is all that new. When Joe McGinniss published *The Selling of the President* in 1968, Americans were shocked by the stories of how Richard Nixon's campaign was stage-managed by Roger Ailes, the young TV producer who would go on to found the Fox News Channel. Forty-seven years later, a Harvard business professor suggested that when the book about the Obama campaign is written, it should be entitled *The Marketing of the President* – as if this was

a novel observation. Even lamenting image politics is an old story.
When Joe McGinniss wrote, "The qualities which now commonly
make a man or woman into a 'nationally advertised' brand are in
fact a category of human emptiness," he was quoting from *The
Image*, a best-selling indictment of mass culture written in 1961 by
Daniel Boorstin.

In all likelihood, politicians have probably always been concerned about how they appear in public, and the public has probably always mocked them for it. But even though Caesar likely got
grief for fussing over his toga whenever he got up to speak, when
it comes to explaining why politics has turned into just another
consumer good, there is no question that shifts in mass communications have had a substantial effect. Radio was revolutionary in
its way, but when it comes to changing the way politics itself is practiced, nothing compares to television. As *Time* magazine columnist
Joe Klein puts it, television "set off a chain reaction that transformed the very nature of politics."

Television altered politics in a number of ways. First, and most
obviously, it made the politician's appearance, his or her image,
and ability to perform in front of the camera their most important
quality. It is frequently suggested that the wheelchair-using
Franklin Roosevelt would not have the slightest hope of being
elected president today, but it is probably more accurate to say that
he would never have been elected governor of New York, let alone
president, if television had existed in the 1930s. It is part of our
culture's received wisdom that it was television that destroyed
Richard Nixon against John F. Kennedy in 1960, beginning the
night a sick and pasty Nixon sweated his way through their first televised debate, his five o'clock shadow pushing through his
makeup. It is widely believed that radio listeners thought Nixon
won the debate while television viewers gave it to Kennedy, though
that remains a matter of some dispute. What is true is that after that
debate Kennedy crept into the lead for the first time, and when he

won the election, it was interpreted as a victory of youth, charisma, and idealism – the quintessential virtues of the television age. As Rick Perlstein puts it in *Nixonland*: "At the ballot box it was almost a tie. On television, in retrospect, it looked as if John F. Kennedy had won in a landslide."

The second thing television brought to politics was money. Lots of it. Before TV, politicians met with their constituents either at town hall meetings or rallies, or indirectly to the masses via the disembodied media of radio or print journalism. Television combined the mass audiences of print and radio with the intimacy of the small group, giving politicians the ability to communicate directly with millions of voters at the same time. But television advertising is extraordinarily expensive, which makes the ability to charm millions of dollars out of the pockets of your supporters yet another vital political asset. In turn, the smell of big money draws the attention of all sorts of political parasites who thrive in the new media ecosystem.

Finally, television affects the way politicians and their spin-doctor courtiers engage one another. As anyone who has ever participated in a televised debate or panel discussion quickly discovers, the medium is an intellectual Flatland, a dominion that abhors nuance, depth, or even the drawing of distinctions. What television thrives on is conflict, and it is important to understand that when political programs invite a liberal and a conservative to debate the issues of the day, the producers have no desire or expectation that there will be any agreement or even concessions by either side. The conflict *is* the message, and while there may be a winner (in the sense that someone might succeed in scoring some prep-school debating team shots), "winning" has nothing to do with being right. The media's pundit class feeds this gladiatorial conception of political debates by treating them as a boxing match, with the postdebate analysis invariably focused on who scored what points, and whether any of the candidates was able to strike the mythical "knockout blow."

The high stakes make debaters increasingly risk averse, and when the presidency or premiership is on the line, the goal is not to win but simply to survive without making any serious blunders. No one wants to be the next Dan Quayle, George H.W. Bush's vice-president, who is remembered for two things: spelling potato with an *e* and comparing himself to John F. Kennedy during his televised debate with Lloyd Bentsen during the 1988 campaign. Bentsen smiled slightly as he gave Quayle the shiv: "Senator, I served with Jack Kennedy. I knew Jack Kennedy. Jack Kennedy was a friend of mine. Senator, you're no Jack Kennedy." It is no wonder then that politics in the age of television consists almost entirely of heavily scripted events that offer little more than prepared statements, canned responses, and memorized talking points. That is why the televised debates between leaders or leadership candidates are so stultifying they make you want to drive nails into your skull.

If anything, the problem is getting worse. Over the past few decades, broadcast network television has been supplemented by the accretion of new media outlets, technologies, and marketing techniques. Cable television brought us the twenty-four-hour news cycle, while new information technology and databases gave us push-polling, market segmentation, and overnight tracking. Now we are seeing the fundamental transformation of journalism itself, thanks to the Internet. The blogosphere turns anyone into an insta-pundit, and a cell phone with a camera makes you into an on-the-spot reporter. With faster computers and cheap editing software, a laptop computer is now a film studio, and with a free YouTube account anyone can quickly disseminate their own attack ads. With the possible exception of the biggest celebrities, politicians spend more time than anyone else in front of a camera, to the point where politics appears thoroughly embedded in what French social critic Guy Debord called "The Society of the Spectacle."

Building on Marx's theory of commodity fetishism, Debord argued that the trajectory of modern society is characterized by a shift from an authentic and active lived reality to an alienated existence that is mediated by images, representations, and passive consumption. The point is not that the consumption of images becomes an increasingly important part of life, but that life itself comes to be lived through representations. As Debord puts it, "The Spectacle is not a collection of images, rather, it is a social relationship between people that is mediated by images."

When you put these three changes together – the focus on image, the role of money and techniques of mass marketing, and the rise of the permanent 24–7 media cycle – it is obvious that critics such as Joe Klein are right: television and its successor media helped transform the nature of politics. Furthermore, there is no question that many people in both North America and Europe find themselves completely turned off by a democratic process that is permanently mired in bullshit (in the strict Frankfurtian sense of the word). What we are looking for, but not getting, from traditional politics is authenticity – a connection to a politician that isn't mediated, marketed, and shaped by how the message will play in the overnight tracking polls.

But there is a problem here, which is that it is the desire for authenticity that leads to a politics that is scripted and sculpted in the first place. We say we want to see the real person behind the mask, but remember the revulsion directed toward Nixon for being all sweaty on television. We say we want spontaneity and real emotion, but think back to the mocking reaction to Hillary Clinton crying on the campaign trail, or the tearful, rambling speech given by North Carolina governor Mark Sanford after he returned from a weeklong tryst in Argentina with his lover. Finally, we say we want to be able to choose from parties and policies that accord with our beliefs and values, but when we are offered a consumer-friendly menu of political "brands," we rebel against the phoniness of it all. In short, the desire for authenticity is the cause of virtually all the major problems with our politics today.

When Stephen Colbert coined the term *truthiness* to describe American political discourse, he defined it as: "What I say is right, and [nothing] anyone else says could possibly be true. It's not only that I *feel* it to be true, but that *I* feel it to be true. There's not only an emotional quality, but there's a selfish quality."

Colbert is bang on. What matters is not the facts, not the truth of the matter, but an emphasis on emotional truth and personal perception. But this is just another way of describing authenticity. The story of how Americans demanded authenticity in their politics and – to their great chagrin – got exactly what they asked for, begins with something called Turnip Day.

Harry Truman had been running a caretaker presidency since the death of Franklin Roosevelt, and nobody expected much from him in 1948. The Democratic Party was divided into three factions, and the Republican candidate, Thomas Dewey, was widely expected to win the upcoming election. But at the Democratic convention confirming his nomination, Truman gave a remarkable speech. Coming on stage after midnight, he spoke plainly, simply, and without notes. He told the audience that he fully expected to win the election, and he challenged the "do-nothing Congress" to get to work.

> On the 26th of July, which out in Missouri we call Turnip Day, I am going to call Congress back and ask them to pass laws to halt rising prices, to meet the housing crisis – which they say they are for in their platform . . . I shall ask them to act upon . . . aid to education, which they say they are for . . . civil rights legislation, which they say they are for . . .

Joe Klein calls this Truman's "Turnip Day" moment, and he loves it. Klein sees Truman's speech as a dose of pure authenticity, in both style and content. It was not a prepared address. No speech-writer had gone over the words and carefully excised anything con-troversial. It was off-the-cuff, earthy, and courageous, especially when it came to Truman's declaration of support for civil rights leg-islation. "In the process," writes Klein, "Truman was able to remind voters who he was – an average guy, a man of the soil, who was plainspoken often to a fault."

Klein wishes that our politicians allowed themselves more Turnip Day moments. He thinks our politics could do with more spontaneity, courage, and authenticity. Instead of talking haircuts, he wants leaders who are comfortable in their skin, who aren't afraid of going off-message or making a mistake, and who are willing to tell the style consultants to take a hike. In the conclud-ing pages of his *Politics Lost*, Klein puts out the call for a leader who can tell a joke, cry, get angry, even indulge in the odd vice or guilty pleasure, within reason. Politicians, he says, need to "figure out new ways to engage and inspire us . . . or maybe just some simple old ways, like saying what they think as plainly as possible."

This is an extremely widespread sentiment. As *New York Times* columnist David Brooks likes to tell readers, ever since the Reagan revolution of 1980, Americans have always voted for the presiden-tial candidate who does the best impression of a fraternity brother, and if you go down the list of pairings over the past eight presi-dential elections, you have to concede that he's on to something. Reagan and Bush the First over Carter, Mondale, and Dukakis; Clinton over Bush and Dole; and then Bush the Second over Gore and Kerry. In every case, the American people opted not for the candidate who offered the best policies, who had the most experi-ence, or who had assembled the most competent team around them. No, they chose as their leader the man they most wanted to have a beer with.

By the time the 2008 presidential election rolled around, the "authenticity" meme had completely taken hold. As a result, competent candidates such as Mitt Romney (who had plenty of executive and private sector experience) and Hillary Clinton (who had spent the last eight years becoming an expert on a number of key files) got steamrolled, as both parties had decided that the only chance they had of winning was to find a candidate who did not suffer from a perception that they were too slick, too phony, too prepared, too stiff, or simply too private. From both Republicans and Democrats the ensuing campaign was an absolute buffet of authenticity, where each candidate offered something a little different in the way of originality, courage, spontaneity, or straight shooting.

On the Republican side there was John McCain, who had a long-standing reputation as one of America's most "authentic" politicians – as mentioned earlier, his campaign bus was called the Straight Talk Express. Authenticity-wise, McCain's counterpart on the Democratic Party ticket was vice-presidential nominee Joe Biden. As the chairman of the Senate Foreign Relations Committee, Biden had become one of the most influential voices in Congress, but his Chiclet teeth and hairplugs gave him a thoroughly prepackaged sheen. Yet given his tendency to speak first and think later, Biden had also earned himself a reputation as a loquacious and somewhat gaffe-prone campaigner.

It was with the selection of Barack Obama as the Democratic Party's candidate for president and John McCain's selection of Sarah Palin as his running mate that things really got interesting. Palin was the moose-hunting maverick governor of Alaska who came down to the lower forty-eight to defend the working-class authenticity of rural and heartland America against the elitism and pretension of the urban elites. Meanwhile, Barack Obama managed to win his party's nomination, and then the presidency itself, despite being a one-term senator with virtually no executive experience. He staked his candidacy on his promise to bridge two of the great divides in American life: the racial divide between blacks and whites, and the social and cultural divide between red states and blue states. Obama's postracial and postpartisan authenticity would take America beyond both the cynical triangulation of the Clinton era and the hardnosed pandering to the party's "base" that characterized the strategy that Karl Rove used to engineer George W. Bush's two victories. Yet as the campaign wore on, Obama found his credibility on both counts called into question.

Few presidential candidates were greeted with as much hope and controversy as Barack Obama, the junior senator from Illinois. On paper, Obama was dream candidate for the Democratic Party. The son of a black father and white mother, Obama was born in Hawaii and educated at Columbia and Harvard. He taught constitutional law for a decade at the University of Chicago before being elected to the Senate in 2004. What made Obama such an appealing candidate was that he was accomplished, charismatic, but best of all, racially mixed. Only one question nagged at his candidacy: is Barack Obama "authentically black"?

The problem had something to do with his mixed-race heritage. Across the political and racial spectrum, Americans cleave to the "one-drop" rule, which holds that a black person is any person with any known African black ancestry. While the rule was used by white racists in the early twentieth century as a way of introducing legislation to prevent miscegenation (and keep the white blood line "pure"), it was later turned on its head by black activists who used the one-drop rule to enlarge the size of the black community in order to enhance its political power and cultural influence. In contrast with people in Latin American countries, where skin color is the subject of a highly politicized color chart (dark = low status, light = high status), African Americans are comparatively easygoing about shades of black. The light-skinned Vanessa Williams is considered no less a member of the black community than, say, the much darker-skinned Wesley Snipes.

No, the problem for Barack Obama is not how much blackness is in his blood, but where that blood is from. It is impossible to separate the American discourse on race from the question of slavery, and almost all of the slaves brought to the United States by white slavers were taken from the curve of West Africa that stretches from Senegal, down through Liberia, as far south as Angola. For many black Americans, especially the older generation that came of age during the flood tide of the civil rights movement, in order to be an authentic African American you have to be descended from those West African slaves. But Obama's father was born in Kenya, which is on the east coast. As civil rights activist and Pentecostal minister Al Sharpton put it, "Just because you're our color doesn't make you our kind."

Once again, it took a fake news program to expose the rank absurdity of this position. Debra Dickerson, columnist and author of *The End of Blackness*, appeared on the satirical *Colbert Report* to assert that Obama is not black because "in the American political context, 'black' means the descendant of West African slaves brought here to labor in the United States." But Colbert, making a point that mainstream journalists were almost uniformly too afraid to make, wondered why Obama didn't simply run as a white man? Or if he was not an authentic American black man, couldn't he run as "nouveau black"? Dickerson found this latter suggestion congenial, noting that as a genuine African American, Obama was a sort of "adopted brother." Colbert went on to tie Dickerson in knots, finally rendering her speechless with the suggestion that perhaps the way for Obama to get some authentic-black cred was for him to spend a year or so as Jesse Jackson's slave.

Things didn't get any easier for Obama after he won the election. Shortly before Christmas 2008, Chip Saltsman made a dumb decision. The former campaign manager for Republican presidential candidate Mike Huckabee was angling for a job as Republican National Committee (RNC) chairman, and by way of sucking up to members of the committee he sent them all a CD of songs that included a song parody called "Barack the Magic Negro." Sung in the cadences of Al Sharpton to the tune of "Puff the Magic Dragon" and popularized by Rush Limbaugh, the reworked lyrics suggested that Obama's popularity stemmed from the fact that he assuages white guilt by making whites feel noble about voting for such a safe and nonthreatening black candidate.

The ensuing uproar effectively torpedoed Saltsman's hopes of winning the chairmanship. In fact, the RNC was so desperate to ease public fears that the party had been taken over by a bunch of closet Klansmen that they gave the job to Michael Steele, a man with a reputation for being both authentically black and a bit of a loose cannon. Shortly after Obama's inauguration, Steele told *The Washington Times* that the Republicans needed a "hip-hop" make-over in order to go after younger voters. "We need to uptick our image with everyone, including one-armed midgets," he said. "We want to convey that the modern-day GOP ['Grand Old Party'] looks like the conservative party that stands on principles. But we want to apply them to urban-suburban hip-hop settings."

Even if you haven't heard the term *magic negro*, you know what it refers to. It is one of the most manufactured and stereotyped roles in film and fiction, that of the wise, sometimes old, occasionally blind, black person who – despite being in a nominally subservient role – uses his special insight or power to save or redeem the white man. Uncle Remus is the classic example, but think also of the role played by Will Smith in the film *The Legend of Bagger Vance*, the Mekhi Phifer role as Eminem's sidekick in *8 Mile*, or the John Coffey character played by Michael Clarke Duncan in *The Green Mile*. The Matrix has two magic negroes, Morpheus and the Oracle. Virtually every character played by Morgan Freeman over the past twenty years is a magic negro; he regularly spends his time either saving white women from serial killers (*Kiss the Girls*) or literally redeeming soulless white men (*Bruce Almighty*). In fact, once you understand the role, you start to see magic negroes everywhere (every cop show ever made has a magic negro police sergeant) along with close relatives the magic Mexican, the magic Chinaman, and the magic mentally or physically disabled person.

But as offensive as it may seem, Chip Saltsman, Rush Limbaugh, and other white Republicans were not the first ones to accuse Obama of being a magic negro. In fact, when he appeared more or less out of nowhere to challenge Hillary Clinton for the Democratic Party nomination, it was the black establishment in America that first accused him of playing off white guilt and acting like the savior of humankind. Indeed, as things evolved for Obama, questions about his racial background took a back seat to deeper worries about his political views, to the point where he was dogged throughout his campaign by two difficult and fundamentally opposing charges. For many blacks, he was simply not black *enough*, hence the magic negro charge.

As if infighting within the black community weren't enough to deal with, Obama also had to deal with suspicion from within his own party that he wasn't sufficiently committed to the more ideological aspects of the Democratic platform. Despite the fact that Republicans tried to paint Obama as the senator with the most liberal voting record in the house, suspicion remained right up to the end that he was far more sympathetic to Republican positions than he was letting on. This was a predictable consequence of his second claim to the authenticity crown – the idea that his promise was not just postracial, but postpartisan as well.

One day during the 2008 election, I arrived at work to find someone had placed a John McCain "Call to Action" figure on my desk. They are still widely available online for only US$13.95, but if you prefer the Sarah Palin model, it is a bit pricier, at $27.95. As for Barack Obama, he has become one of the great Warholian figures of our age, his image emblazoned on everything from T-shirts and base-ball caps to a delightfully short dress by designer Jean-Charles de Castelbajac, which was exhibited on the runway during Paris Fashion Week. There is also Obama sushi, Obama snow sculpture, Obama sneakers, and a colossal amount of Obama-themed art portraying him as everyone from Jesus to Superman – most of it lovingly catalogued on www.obamaartreport.com.

In Canada (where there was a federal election running parallel to the one in the United States), our prime minister took to wearing a blue sweater vest in the hope that it would change his image from aloof, laser-eyed alien to folksy family man, while the left-wing New Democratic Party, led by a bald, mustachioed man named Jack Layton, fought back against Layton's "Taliban Jack" reputation (he advocated negotiating with the Taliban) by trying to rebrand its position "The New Strong." The fact is, whatever your partisan leanings, there is a brand for you. Politics seems more and more a branch of marketing, with parties and leaders packaged and sold using the same techniques used to sell energy drinks, NBA players, and everything in between.

A lot of people find this highly objectionable. At best, the con-sequence of turning politics into a consumer good is the Big-Macification of civil discourse, where politicians are forced to pander to the lowest common denominator of the electorate. At worst, it supposedly turns the electorate into "Manchurian voters," where the manipulation and propaganda that goes on in political advertising basically tricks or bamboozles people into voting for a sunny image that masks a sinister agenda.

This view of party politics as a distasteful introduction of the techniques of consumer marketing into our democracy is what formed the basis of Obama's postpartisan appeal. It also happens to be completely misguided. The fact is, the selling of politics does not undermine democracy, it enhances it, and the branding of political parties and leaders is not a tool for manipulating voters, but a mechanism for enabling democratic participation.

Consider how brands work. The central question that every consumer faces is, "How do I know I'm not getting ripped off?" How do you know that this bag of flour isn't adulterated or that these new shoes won't fall apart the minute you get home? Unless you've managed to follow the entire production process from start to finish, you don't. You trust the flour isn't full of sawdust because Robin Hood says so. You have faith the sneakers will withstand a running season or two because Nike has put its swoosh on them. Brands are one of the earliest and most effective forms of consumer protection, where trust in the brand (and the company behind it) substitutes for first-hand knowledge or expertise.

Political brands work the same way. In an election, the question every voter needs an answer to is, "How do I know what I'm buying into with my vote? How do I know I'm not getting snookered?" This is where political brands, better known as parties, come in. The role of the party is more or less to take the dense convolutions of modern governance and reduce them to a relatively simple brand proposition. Are you generally in favor of a strong central government that will build national social programs? Then vote Democrat (or, in Canada, Liberal). Would you prefer a more decentralized federation and limited state interference in your life and in the economy? Then the Republicans or Conservatives are the party for you.

The paradox of all branding is that the more complicated things get, the simpler the messaging has to be, which is why politics has become so intensely focused on the party leader's character and image. It's pretty remarkable that in an election in which American voters were being asked to decide who would control a budget of somewhere north of $3 trillion, they were essentially offered a choice between two brands: Barack Obama's "Change" and John McCain's "Honor." But what is more surprising still is how well the system actually works. Most people don't have the time or, frankly, the ability to properly digest budgets, policy documents, or drafts of new bills, and the distillation of the stupendous complexities of the modern state to a handful of simple but distinct brands is not just useful, but necessary. As in the consumer economy so in modern politics – both would grind to a halt without brands as a lubricant.

What of the worry that politics ends up being marketed like Big Macs, pitched to the lowest common denominator? The proper reply is to this is, So what? People always put the emphasis in that phrase on the word *lowest*, when it should be placed on the word *common*. The government wields a monopoly over the use of violence, among other things, and any party that wants to claim the right to use violence had darn well better make sure it has the lowest common denominator onside or it is in big trouble. To adapt a line from the genius of twentieth-century advertising, David Ogilvy: the lowest common denominator is not a fool, she is your neighbor. In a democracy, every politician is in the business of selling electoral Big Macs, and anyone who thinks that's not his job is either a born loser or a tyrant *manqué*.

We need to give voters a little more credit. People are no more bamboozled by a John McCain action figure into voting for John McCain than they are tricked into buying a PC because Jerry Seinfeld is in the ad. That just isn't the way branding or human behavior works. Indeed, whether it's Nike's swoosh, or the ubiquitous Hope poster of Obama designed by Shep Fairey, or Stephen Harper's blue sweater vest, no one ever admits to being a dupe of the marketing. The worry is always that *other* people – in particular, the people who support the other side – are being manipulated. And so throughout the Bush years, the left in America complained about the way Karl Rove and Dick Cheney were sowing fear and panic over terrorism and keeping the religious right all a-boil over fears about abortion and Mexican immigrants. Once Obama became president and the Democrats took control of both houses of Congress, the right immediately started complaining that the electorate had been duped by his pretty speechifying and his wispy promises about Hope and Change.

This is a slippery slope, and it is dangerous for anyone, no matter what their partisan allegiances, to have so much contempt for voters. Democracy is based on the premise that reasonable people can disagree over issues of fundamental importance, from abortion and gay rights to the proper balance between freedom and security. When the mere fact that someone supports the other side becomes evidence that they have been brainwashed, then the truth is you no longer believe in democracy.

During the 2008 campaign, there was at least one person in America who wasn't buying into Obama's postpartisan authenticity shtick, and that was Sarah Palin. When she arrived on the scene at the end of August, she promptly reminded everyone that it was possible to be both partisan *and* authentic. The core of Palin's appeal was a fairly straightforward defense of small-town America and its small-town values of "honesty, sincerity, and dignity," as against the (presumably) dishonest, insincere, and undignified city-dwellers. As she put it in her galvanizing address at the Republican Party convention in early September 2008, small-town people are "the ones who do some of the hardest work in America, who grow our food and run our factories and fight our wars."

There was nothing really new in this – the populist defense of the working classes as being more authentic than the effete snobs and fakers who live in places like San Francisco, New York, and Washington is as old as the hills that it romanticizes, and it forms the basis of the appeal of singers such as Bruce Springsteen and John Mellencamp. The nadir of this brand of blue-collar romanticism was the embarrassment that was Samuel Joseph Wurzelbacher, a former plumbing-contractor turned conservative-pundit better known as Joe the Plumber. Wurzelbacher came to fame during the 2008 election when he confronted Obama about his business tax policy. While he refused to publicly endorse either McCain or Obama, Wurzelbacher repeatedly referred to Obama's plans as "socialism," and he questioned Obama's loyalty to both America and Israel. After the election, he made an ill-advised trip to Israel as a war correspondent, and then quickly faded from public view. What was interesting about Palin's authenticity gambit is the way she used her candidacy to launch America right back into the running battle over religion, social values, and patriotism that has convulsed the nation's politics for the past forty years.

Palin chose to fight this fresh skirmish in the culture wars on some rather odd terrain, with the lines of opposition forming not along traditional issues such as abortion or gay marriage, but, most curiously, over who has the better hobbies. To put it bluntly: Sarah Palin must be the first candidate for vice-president to have staked her claim to the office largely on the fact that in her spare time she enjoys shooting and dismembering large, defenseless mammals. She only alluded to the hobby in her convention speech, but before John McCain's own address the next day they played a video biography, obviously put together in great haste, that was designed to introduce this complete unknown from Alaska to the party faithful. What was the thrust of the message? As the cheesy, movie-trailer voice-over intoned at the beginning of the tape: "Mother . . . Moose hunter . . . Maverick."

All politics is to some extent personal, but Palin seemed determined to outdo both Obama and McCain in making an argument out of her lifestyle. In her speech, she talked about herself a lot, lobbed some insults at city folks, and then talked about her entire family a lot more. She's a hockey mom and her husband works in the oil fields and fishes and races snowmobiles on his days off. Son Track was off killing Iraqis, while daughter Bristol, only sixteen, had gone and done her best Jamie Lynn Spears impression and got pregnant.

When she finally got around to talking about her Democratic opponent, Barack Obama must have felt a bit like a moose caught in her gunsights. With calm precision, she took aim at his reputation as a "community organizer," at his "high-flown speech-making," even at the faux-Greek columns that flanked his own acceptance speech.

What Palin did was something the media (and most Republicans, for that matter) had not had the temerity to do, and that is refuse to give Obama the deference owed to his status as the first black candidate for president. Instead of respecting his background, she made fun of it, treating him as just another too-thin, Ivy-League-educated, wine-sipping member of the liberal elite. Perhaps more than anyone else, she accepted his "postracial" branding, treating the fact that he happens to be (half) black as irrelevant. In choosing to pick on Obama's lifestyle, Palin struck closer to the new realities of the supposed culture war than even she seemed to realize.

In his 2004 book *What's the Matter with Kansas*, critic Thomas Frank complained that for decades Republicans have been effectively conning the party's base. They get the supporters all heated up over things like gay marriage and abortion during campaigns, but once in power they quietly put those issues on the back burner, turning their attention to issues that matter to their fiscal conservative backers: cutting taxes, freeing trade, and eviscerating the welfare state. Through this massive electoral bait-and-switch, Frank argues, good-hearted authentic folk from the heartland are bamboozled into voting against their economic self-interest. Always willing to trade economic hope for religious comfort, they serve as cannon fodder in a culture war the Republican elites have no intention of trying to win.

Frank's argument is actually just a variation on Obama's own ill-advised remark about "bitterness" causing people to cling to guns and religion. Obama and Frank think the sorts of people who like to thump Bibles and shoot moose and ride snowmobiles also tend to be economically disadvantaged: that is, they assume that people with working-class values must also have working-class incomes, and that the culture war is actually a disguised class war between well-off social liberals and not-so-well-off social conservatives.

This is an assumption that Palin herself was happy to play off in her attacks on Obama. But if it was true, once upon a time, that you could tell how much someone earned by how they spend their spare time, it no longer is. Hunting and fishing, RVing, camping and canoeing, dogsled and snowmobile racing – these are now the working-class pursuits of rather well-off people. A canoe trip on a river in the Northwest will set you back $6,000 or $7,000, while a week at a salmon lodge on the Restigouche River in New Brunswick starts at around $10,000. And has anyone priced a racing-quality snowmobile recently, not to mention a class-A motorhome? In comparison, the stereotypically urban fondness for egg-white omelets and yoga classes looks downright frugal.

In the end, the candidacy of Sarah Palin didn't herald a re-engagement of the culture wars but their exhaustion. What forty years of arguing over authenticity in American politics has boiled down to is a disagreement over who has better leisure pursuits. It's no longer a dispute over fundamental values, a fight to the death for the soul of America. Instead, Sarah Palin turned it into a dinner-party argument about who has better taste.

Nothing more perfectly characterized Palin's impact on the American political scene than her leaving of it, however temporary her departure may turn out to be. When she abruptly resigned as governor of Alaska just halfway through her first term, there was a great deal of speculation about her motives. Was she tired of the endless attacks on her family? Was she the target of a secret FBI investigation? Or was she merely stepping back from the fray in order to get a running leap at the presidency in 2012?

Her behavior at the press conference in which she announced she was stepping down didn't do much except confirm that when it comes to speaking her mind, she makes men such as Biden and McCain look like models of discretion and sober second thought. Her speech was a meandering rant of bitterness and indignation, aimed at anyone who ever drank a cappuccino, looked down at Alaska, or worked as a journalist. When it came to the only thing anyone really wanted to know – why she was quitting – she went off on some gonzo basketball analogy about how a good point guard knows when to protect the ball and when to pass it off for the good of the team.

At that moment, it became clear that Sarah Palin wasn't simply making things up as she went along. Instead, it would seem that she's as much a bystander to her own consciousness as the rest of us; stuff just comes out of her mouth and she lets other people figure out what to make of it. Whatever it is, it sure was spontaneous. But as Dahlia Lithwick wrote in a column for *Slate* magazine, the problem with this approach is that while "it's all well and good to be mavericky with one's policies, it's never smart to be mavericky with one's message."

We seem to have reached an impasse. On the one hand, many of us find ourselves alienated from a political system that is so tightly messaged that all of the life seems to have been squeezed out of it. Yet at the same time, the candidacy of someone such as Sarah Palin is the *reductio ad absurdum* of the conceit that the simple willingness to speak one's mind is what people mean when they say they want more Turnip Day moments from their leaders.

Begin with the obvious point, which is that nobody sets out to do "fake politics." Every rising star in the political heavens arrives on the promise to "do politics differently," which is code for "I won't lie, dissemble, waffle, temporize, flip-flop, or otherwise mislead you." The promise is that they will do politics differently by being direct, honest, and courageous and by simply being themselves. The pledge to do politics differently usually lasts for about a month, until an off-the-cuff remark or straight ahead answer to a question is twisted by the media into a "gaffe." (As the saying goes, in politics a gaffe is when someone tells the truth.) At this point, the politician either learns to survive by mastering the fine art of laying smoke, or he or she crashes to Earth, dismissed by the public as an error-prone buffoon or a dangerous wingnut.

In truth, a great share of the blame lies with the media and its obsession with controversy and scandal at the expense of more difficult question of policy and other serious issues. The press lies in wait, ready to pounce on anything that can be labeled a gaffe, which causes their prey to stick ever closer to the protection of their talking points. The politicians end up looking more and more like brainless automata, which leads the public to treat them with increasing contempt. The fact is, politicians and the press are locked in a Mexican standoff. It is a race to the bottom that no one can win, but neither side has any incentive to stop.

The second difficulty has to do with the attitudes of the public. Do we want a leader to be himself? When Stephen Harper took over as prime minister of Canada in 2006, he was a fashion disaster. At one early press conference he wore a gold golf shirt that showed off his jiggling man boobs, photographs of which were widely circulated and published in the satirical press. On an official visit to Mexico, he wore a khaki fishing vest that made him look like George W. Bush and Vicente Fox's dim-witted little brother. Harper became an object of ridicule for being himself, and there are few men out there who could withstand that much heckling and *not* turn to an image consultant for help.

Do we want spontaneity? While campaigning in the spring of 2007, John McCain was asked if he had any ideas for how to deal with Iran and its refusal to stop enriching uranium. He picked up the microphone and immediately broke into song, singing "Bomb Iran" to the tune of the Beach Boys hit "Barbara Ann." His audience laughed, but as the incident rippled through the blogosphere, it only reinforced the widespread belief that he was a bit too crazy to be president.

The most damning example of our two-faced attitude toward spontaneity, though, is the stunning rise and incredible flameout of Howard Dean. A six-term governor of Vermont, Dean ran for the Democratic Party nomination for president in 2004. He advocated a "50 State Strategy" that would see the Democratic Party fighting for votes in traditional Republican strongholds, and he pioneered the online fundraising model that Obama would use so successfully four years later. He was an early leader in the race, and for a while it looked like he might run away with the nomination. Until, that is, an unfortunate night in Iowa, when Dean finished a disappointing third in the state's caucuses, behind John Kerry and John Edwards.

At an evening rally for his supporters, Dean gave a speech in which he vowed to carry on the fight. As he got increasingly worked up, his face got red and his throat tightened, until he started yelling about how "we're going to South Carolina and Oklahoma and Arizona and North Dakota and New Mexico, and we're going to California and Texas and New York . . . And we're going to South Dakota and Oregon and Washington and Michigan, and then we're going to Washington, D.C., to take back the White House! Yeah!!!"

That final "yeah!" sounded like a pig being strangled, and it punctuated what Dean himself later conceded was "a crazy, red-faced rant." A clip of the speech was played repeatedly on TV, and then it was posted on YouTube, where it was remixed and repackaged as "The Dean Scream." Dean hung on for another month, but his candidacy effectively ended that night, his credibility as a man in control of his emotions completely shot.

The truth is, we say we desire more authenticity in our politics, but when push comes to shove, we want authenticity only when it mirrors our own narrow values and ideals. The reason politicians hire image consultants and stick doggedly to their talking points is that spontaneity and frank talk is punished far more frequently than it is rewarded. That is why it is the public, not the spin doctors or the media, who are to blame for a political culture that is as bland and homogeneous as fast food and DIY furniture. Like McDonald's or Ikea, political leaders are trying to appeal to as many people as possible without turning anyone off. Worse, they have the additional burden of trying to do it under the omnipresent and omni-hostile gaze of a media that will rip them apart at the slightest misstep. Given the alternatives, it is easy to see how a $400 haircut and full-time image consultant may start to look like a bargain.

During the 1993 Canadian federal election, the incumbent Progressive Conservative Party ran a series of attack ads against Jean Chrétien, the leader of the opposition Liberals. The second ad consisted of a number of closeup photographs of Chrétien's face, his face twisted in an apparent sneer, with one side of his mouth open wide, the other almost shut. A succession of voice-overs criticized Chrétien's policies but also asked rhetorical questions such as, "Is this a prime minister?" Most of it was fair game by the relatively unsporting rules of politics, but the last comment in the ad – "I would be very embarrassed if he became prime minister of Canada" – caused problems. The photos used in the ad weren't just bad candids; Chrétien does have a slight facial deformity, the result of Bell's palsy, which he has had since he was a teenager. It becomes more prominent when he is speaking loudly.

There was an immediate backlash against this form of "negative campaigning," and Prime Minister Kim Campbell immediately ordered the ad off the air. Her handlers complained that the backlash was largely a media creation, but the damage was done. As one pundit put it, Chrétien went on to make the speech he had been waiting his entire career to deliver, telling a Nova Scotia audience: "When I was a kid people were laughing at me. But I accepted that because God gave me other qualities and I'm grateful." Members

of the audience had tears in their eyes. After nine years of tumultuous rule, the Tories were always going to have a hard time retain-
ing power, even with a new leader. But after the fiasco over the
attack ad against Jean Chrétien, the party never recovered in the
polls and the Liberals went on to win in a landslide. The former
government was almost destroyed as a party, reduced to only two
seats in a 295-member House of Commons.

This affair appears to illustrate the widespread conviction that
the public does not approve of attack ads and will punish parties
who are deemed to have crossed the line of good taste and fair play.
One of the eternal verities of elections is that, at the beginning of
every campaign, each party solemnly swears that they intend to
run a clean, positive campaign. And if they end up in the gutter, it
is because their opponent dragged them down into it. "Going negative" is usually interpreted as a sign that a campaign is in trouble,
a desperate gamble that if everyone ends up covered in mud, the
public won't really care who threw the first handful.

Almost every losing party or candidate eventually gets desperate, which may explain why we have some negative campaigning.
But that does not explain why there is so much of it or why attack
ads make up an increasingly large proportion of all campaign
advertising. According to a study by Brown University professor
Darrell West, negative advertising in U.S. presidential elections
climbed steeply in the 1970s. In 1976, 35 per cent of all major ads
were negative, rising to 60 per cent in 1980, 74 per cent in 1984,
and 83 per cent in 1988. That was the high-water mark until 2004,
when the Bush–Kerry mudfight set new standards for negativity.

So the public strongly dislikes attack ads, and politicians themselves say they would rather not run them. Furthermore, they are
widely blamed by academics and journalists for lowering the level
of civility in public discourse, hurting voter turnout, and alienating
citizens from their elected representatives and democratic institutions. So why is there so much of it about?

The straightforward answer, as any political strategist will tell you, is that they work. "Positive" ads, which are usually warm and fuzzy spots talking about how the sun will shine brighter and the birds will sing sweeter only if Pat Smith gets elected, are generally ignored by a skeptical public. But when Chris Jones criticizes the record or impugns the character of Pat Smith, people stop and pay attention. As one veteran political consultant argues, "People always say they hate it when newspapers print photographs of car crashes, too. And then, when they're near a car crash, the same folks slow down to take a good look." Even the man responsible for the notorious ad making fun of Jean Chrétien's facial deformity agrees. Allan Gregg, a pollster and market research consultant in Toronto, continues to insist that the ad would have worked if it had been allowed to run longer. Gregg thinks the uproar was largely a creation of the media, and it wasn't the ad that caused the damage but Kim Campbell's decision to "wear" the backlash by pulling the ad off the air.

The fact that they work would certainly explain the prevalence (if not the popularity) of negative political advertising, but this only pushes the question back a step. Why do they work? Part of the answer has to do with some of the peculiar ways in which political marketing differs from conventional commercial advertising. But odd as it may sound, much of the blame must lie with our very desire for more Turnip Day moments from our politicians, because the murky depths of "authenticity" is where the character assassins of negative advertising make their home.

A lot of political advertising gets labeled negative when it isn't. A study by the Annenberg Campaign Mapping Project usefully sorts political messaging into three types: advocacy messages, which give arguments in favor of a politician's position; contrast messages, which contrast two or more positions or choices; and attack ads, which are straightforwardly critical of an opponent's position or character.

One of the more jarring features of political advertising is how different it is from the more familiar commercial variety. The identity of a strong commercial brand, such as Apple, Coca-Cola, or Volvo, is established almost entirely through positive or "advocacy" marketing. The brand's identity, or "unique selling proposition," is created through spots that deliver a clear and consistent message about the product's position in the marketplace. Every Coca-Cola ad builds on the idea that Coke embodies authenticity, and each Volvo campaign supports the promise that their cars mean safety. Overwhelmingly, commercial advertising sticks to the rule that you never mention the competition: Sprite doesn't run spots showing a kid at a birthday party taking a sip of 7-Up, then promptly barfing all over the cake. Adidas does not have a campaign that features cops chasing violent criminals through the streets, only to come up short because they couldn't run fast enough in their Nikes.

In fact, if you ask people to think of negative commercial advertising, almost everyone mentions the Pepsi Challenge (if they can remember that far back), or another more recent example, Apple's Get a Mac campaign, which shows a hip, casual, laidback Mac guy interacting with a stuffy, uptight, work-obsessed PC person. But neither of these are genuine negative or attack ads; they are actually contrastive spots that promote the merits of one brand over another. True negative advertising, where one player in an industry goes flat out against a competitor, is almost unheard of in the commercial realm.

One major reason is the sheer number of competitors. Down at the food court, you can choose from McDonald's, Burger King, Harvey's, or Wendy's, and that's just if you are in the mood for a burger. So if McDonald's runs a negative ad against Burger King, it could give customers a reason to avoid the Whopper, but they may very well head over to A&W for a Mama Burger. Meanwhile, in North American politics at least, the only serious choice open to voters is Democratic or Republican in the United States, Liberal or Conservative in Canada, and a reason for not voting for one is implicitly a reason for voting for the other. It is like in church, where the choice is between either God or Satan. It is all well and good to preach to the choir about the glory of God, but it also helps to occasionally remind the flock about what a bad guy the devil is.

What's distinctive about the commercial realm is that the size of the market is not fixed, and a successful brand can actually create a niche in which many competitors can flourish. In fact, one of the "immutable laws" of branding is that a leading brand should promote the product or service category as a whole. The idea is that if you build up overall market awareness, the brand leader will benefit disproportionately – think of what Starbucks did for coffee or what Red Bull has done for energy drinks. The corollary of this explains why there's so little negative advertising in the shopping mall. Why doesn't Kenneth Cole go after Ralph Lauren? Because it would run the risk of turning the public off the entire category and shrinking sales for all concerned.

In contrast, politics is a business where power is the only spoils. And since you either have power or you have nothing, it means that, unlike the market for coffee or burgers, politics is a zero-sum game. All that matters is winning, and there is no risk of shrinking the entire market by running down your competitors. Under these conditions, negative advertising becomes an absolutely necessary weapon in your communications arsenal.

Finally, another significant difference between politics and commerce is that while the fight for market share in the supermarket is a ceaseless battle, politics consists of long periods of monopoly power punctuated by short bouts of heavyweight competition. If the supermarkets were to periodically hold a six-month long contest during which consumers would vote on which cola would own the supermarkets of the nation for the next four years, you would see a heck of a lot more negative campaigning. In particular, the loser of the previous "election" would probably spend most of the campaign criticizing the taste of the winning cola and denouncing the corporate owner for being lazy and arrogant and taking the consumer for granted.

As for the notion that negative advertising is a poisoned chalice that in the end only harms democracy, this is certainly the received wisdom about negative campaigning. In *Politics Lost*, Joe Klein laments the influence of the Democratic strategist Patrick Caddell, a genius of negative campaigning who came to "the perverse realization that he could make the race so obnoxious that he would actively discourage people from voting."

This is a depressing observation, but also perhaps a mistaken one. For one thing, studies of the content of political advertising in the United States show that negative ads actually have more factual content, and deal with more policy issues, than other forms of communication. But more interesting still is new research that has found that the more people dislike the candidates and supporters of the other parties, the more likely they are to vote. This suggests that, if anything, negative campaigning should increase voter turnout.

But is voter turnout such a good measure of the health of a country's democracy? If history is any guide, there is such a thing as a society that is *too* politically engaged. In Weimar-era Germany, virtually every institution of civil society, from gyms to music groups to outing clubs was organized along partisan lines. In this overheated democracy, where voter turnout in even the most minor elections was routinely more than 80 per cent, Germans became accustomed to the idea that just about everything you did in the public sphere was at least an implicit reflection of your political leanings. The fact that German society was already so heavily politicized made it that much easier for the Nazis to "coordinate" these institutions once they gained power, because it wasn't so much a matter of introducing politics into civil society, as simply Nazifying institutions that were already politically compromised.

So the prevalence of attack ads has something to do with both the nature of the spoils and the incentive structure of the political marketplace, but what probably exerts an even greater influence is the very desire for more authenticity in our politics. Remember that the supposed virtue of Turnip Day moments is that they reveal gaps in the armor, which afford a window into a politician's true self. It is after the speechwriters and handlers have gone to bed, when they have had a few drinks or are really wound up, that politicians allow themselves those moments of spontaneity that reveal their real character, their deeper humanity.

This assumes that "humanity" is somehow self-revealing, that you know it when you see it. But like the ad man said, if you can fake sincerity, you've got it made, and the two most successful American politicians of the past thirty years, Ronald Reagan and Bill Clinton, were in a league of their own when it came to faking sincerity. Worse, even if character *were* self-revealing, nothing says we are going to like what we see. Some of us have a heart of gold, others a heart of darkness, and once we decide that character is a chief reason for voting for one candidate and rejecting another, the table is set for one hell of a food fight. If character matters, then the moral valence of that character matters, and it becomes a legitimate target of attack ads. In fact, once authenticity becomes a much-prized quality in our leaders, attack ads become not just likely, but obligatory. For example, during his 2008 primaries battle with Hillary Clinton, Barack Obama was caught cribbing a few lines for a speech from his friend Deval Patrick. Clinton was quick to pounce, arguing that this "plagiarism" was no mere dishonesty; it actually undermined "the entire premise of his candidacy." Clinton had a point: since plagiarism is passing off the work of another as your own, it is a crime against authenticity, and authenticity was a major element of Obama's brand.

The same danger awaits anyone who rests his or her case for public office on their character. If a Rudy Giuliani–type of politician chooses to run for president on his reputation as a man of strength and integrity, then the fact that he and his wife have six marriages between them becomes a legitimate matter of public interest and concern. If a John Kerry decides to base his entire campaign on his service in Vietnam, then the nature of that service, and the character he exhibited while in-country, is something that has to be critically examined.

This fixation on authenticity explains why "hypocrisy" has become the greatest political vice. A political hypocrite is someone who supports one set of moral rules or principles that should apply to the general public, while adhering to a separate (and usually more lax) moral code in their private life. For obvious reasons, social conservatives are most at risk here, and when it was revealed that William Bennett (author of numerous books on America's supposed moral collapse) had a long-standing gambling addiction, the liberal press and public delighted in having unmasked him as a hypocrite.

Unfortunately, this puts politics into a death spiral. A fixation on authenticity and a candidate's character creates an opening for attack ads by the opposition. But this in turn gives a candidate an incentive to lie about his past or hide his true character, which provides jobs for all the spin doctors and image consultants whom nobody likes. In the end, Joe Klein has it exactly backward: it isn't the spin doctors who have drained the authenticity from politics; rather, it is the desire for authenticity that provides opportunities for men who can help you fake it. The only alternative is to vote only for candidates who are so upright, honest, and unimpeachably dull that you wouldn't want them having supper with you, let alone running the country.

—

SEVEN

CULTURE IS FOR TOURISTS

ON THE WESTERN EDGE OF DOWNTOWN MEXICO CITY SITS Chapultepec Park, a 988-acre oasis of trees, playing fields, and gardens that rivals New York's Central Park as one of the world's outstanding urban playgrounds. And one of the great attractions in Chapultepec itself is the Museo Nacional de Antropología, an enormous museum that contains a seemingly endless succession of exhibition halls, each dedicated to a separate period or culture in Mesoamerican history. Just off the large open plaza that dominates the entrance to the museum is a dirt-covered clearing ringed with benches. At the center is a 65-foot-high metal pole with a platform on top, like a ship's mainmast without the cross-trees. This clearing is where tourists gather to see the dance of the Voladores, an ancient pagan rite performed by Totonac Indians from the Papantla region of Mexico, in Veracruz.

For the performance, five men dressed in brightly colored traditional costumes climb to the top of the pole. Four of them tie to their ankles thick ropes that have been wound around the top of the pole, then fling themselves off headfirst and backwards, like scuba divers. As the ropes unwind, the four Voladores spiral to the ground in slowly expanding circles, while the leader of the group, known as the *caporal*, plays a drum, a flute, and prays to the fertility gods. While all of this is going on, a handful of assistants – clad in the same traditional getup – canvass the crowd for donations "to the gods," and maybe a bit extra for the brave men who have performed the dance.

What distinguishes the dance of the Voladores from routine busking is its purported cultural authenticity. Although it seems to have survived largely as it was reported by the earliest Spanish colonists, no one knows for sure the origins or full significance of the ritual. This is partly because the Spaniards made a point of destroying all of the indigenous documentation, but also because these same Spaniards were quite sure that what they were seeing was not a religious ceremony but some sort of sport. The upshot, anyway, is that the dance of the Voladores is a living artifact, a museum piece as frozen and uncertain as the masks and figurines and objects that fill the Museo Nacional de Antropología itself.

This is far from an isolated phenomenon. Just about every place worth visiting makes a point of promoting a preserved form of its supposedly pure and undiluted cultural past to tourists and other visitors. Often it involves Aboriginal groups – see the live singing and drumming by the Cowichan people on Vancouver Island, or watch the Maori dance around in body paint and traditional clothes in New Zealand. But you can also go to resorts in the Caribbean where they all dance around with fruit on their head even though you know darn well that no one carries fruit on their head in the city. Or you can visit the Jewish quarter in Krakow to drink kosher vodka and listen to Klezmer music played by university students from Toronto. The accusation that has been leveled against this sort of cultural preservationism is that it comes at the cost of turning a living tradition into a museum piece. As a Pacific Island dancer replied when asked about his culture: "Culture? That's what we do for tourists."

The idea that cultural authenticity is something fit only for tourists is the logical consequence of the idea that a traditional society is meant to be closed, particular, and internally homogeneous. What has been lost is the way any living culture has to be open to and engaged with the world. Philosopher Dennis Dutton illustrates this with a nice thought experiment.

Begin by imagining the dense and interlocking layers of talents, abilities, stores of knowledge, techniques, traditions, and so on that make up the art of opera as presented by a great company such as La Scala. And consider also the way that opera is surrounded by a world of criticism, scholarship, and historical understanding – that is to say, an audience – that makes the opera a living critical tradition.

What would happen then if one day La Scala were to lose its natural, indigenous audience? Italians and other Europeans stop going, local newspapers stop reviewing new performances. Instead, it becomes a destination almost exclusively for tourists, for whom the La Scala opera is maybe the first and even last opera they will ever attend. It is for them the cultural equivalent of kissing the Blarney Stone or visiting the Grand Canyon, and "although they are impressed by the opulent costumes, dazzling stage-settings, massed chorus scenes, and sopranos who can sing very high, they cannot make the sophisticated artistic discriminations that we would associate with traditional La Scala audiences of the nineteenth and twentieth centuries." As Dutton points out, the opera still has an audience, but it has lost its connection to an evolving critical culture. Indeed, the nominal culture of origin might eventually forget where the art form came from in the first place, and lose track of what the signs and symbolic elements of the form mean.

This is pretty much what has happened to the Voladores of Papantla. And not only to the Voladores, but to the countless rituals, ceremonies, traditions, and even entire cultures that have been preserved for display, in all their authentic glory. But at what price authenticity? These have been preserved only in the most literal sense: the crowds come to gawk and gape, but like the thousands each day who shuffle past Mao's embalmed corpse in his mausoleum at the southern end of Tiananmen Square, it's context-free amusement.

The closer you look at it, the more the very idea of authentic culture seems to drift away into incoherence. In a nice metaphor from Kwame Anthony Appiah, a Ghanian scholar who teaches at Princeton, examining a culture is like peeling an onion, where you discover layer upon layer of influences, borrowings, re-imaginings, and wholesale imports from other places. As an example, he points out that in West Africa, traditional Herero dress for women comes from nineteenth-century Lutheran missionaries. In Canada, it is customary for political leaders to give as gifts to visiting dignitaries Inuit soapstone carvings, but few Canadians realize that carving was introduced to the Inuit by a white carver in 1948.

Examples like this are endless. Every aspect of almost every culture, from musing to music, from dining to dance and everything else you can think of, has been shaped by trade in goods, ideas, technologies, and – more than anything else – by the simple fact of people moving around the planet and interacting with one another. One of my favorite examples is the steel drum ensembles of Trinidad, whose main instruments are the fifty-gallon oil barrels left behind on the island by U.S. forces after the Second World War. These drums, along with other metallic objects such as biscuit tins and frying pans, almost completely replaced the indigenous drum technology, which used bamboo. But does anyone think Trinidadian steel drum music is any less "authentic" for it?

A healthy culture is like a healthy person: it is constantly changing, growing, and evolving, yet something persists through these changes, a ballast that keeps it upright and recognizable no matter how much it is buffeted by the transformative winds of trade. We can even expand the analogy a bit, and think of a culture as something akin to a society's immune system – it works best when it is exposed to as many "foreign" bodies as possible. Like kids raised in too-clean environments, cultures that are isolated from the world are like porcelain dolls – beautiful but extremely fragile. That is why, when it comes to protecting and preserving the particular cultures of the world, "authenticity" of the sort that natives engage in for tourists is probably the last thing we should be concerned with. A more appropriate focus of our concern is something more flexible, engaged, and robust, which we can call a "worldview" or "ethos."

Before it was a brand of bottled water sold at Starbucks, *ethos* was a useful Greek term that refers to the moral, political, religious, artistic, and scientific character of a society. It is the background set of customs and assumptions, the rituals and symbols, the rules and hierarchies that determine how and on what terms a society engages with the world and with other peoples. Ethos is more or less what most of us are getting at when we talk about a cultural identity. It is a unique perspective on the world, or, simply, "worldview." This identity or worldview is what motivates a society's artistic creations and its scientific innovations, and if it is sufficiently self-confident in its ethos (as in ancient Greece, or Renaissance Florence, or mid-twentieth-century America) it can achieve remarkable things.

In short, a strong cultural ethos or identity is what sustains almost everything that makes life worth living – as distinct from the sorts of things (food, shelter, medicine) that make life merely possible. And this is a problem. Because while globalization is very good at bringing, through trade, the spread of ideas and the development of new technologies and better and cheaper versions of life's needs, as often as not globalization has a deleterious effect on particular forms of culture.

Sometimes even the mere awareness of the outside world can destroy a culture by undermining its self-confidence and overturning established hierarchies and institutions. This is the essential conceit of the Prime Directive in *Star Trek*, as well as the basis for films such as M. Night Shyamalan's *The Village*. A more prosaic but in some ways more instructive example is the manner in which an artist's vision or voice changes, frequently for the worse, when he or she becomes successful. This is a problem that affects many writers, musicians, and filmmakers, and it is commonly referred to as the sophomore slump. Many writers (Joseph Heller, Emily Brontë) have only one good book in them, and many successful bands struggle to put out a decent second album.

Drugs and success-induced lassitude aside, the sophomore slump is the result of the subtle but inevitable transformation of an artist's postsuccess worldview or identity. They become self-aware and self-conscious, and the original spontaneity, innocence, and naturalness that gave their early work so much confidence becomes more calculated, jaded, and ironic. The artist may fall victim to the anxiety of influence, the sense that everything worth saying has been said before, every song has been written, and that all that remains of creativity is footnote and quotation.

As with an individual artist, so with an entire society. For all the benefits that the outside world can bring, it is possible for a society to be overrun by external influences, diluting and undermining the culture to the point where little remains of the original ethos. The possible destruction of a unique and probably irreplaceable worldview is one of the real and tragic consequences of globalization. What makes it all the more heartbreaking is that the forces that lead to the destruction are both thoroughly legitimate and almost impossible to control or impede in a free and open society. Individuals have the desire and the right to improve their lives through trade, to adopt new technologies, to explore other ways of seeing the world. This process by which individuals acting in their own interest collectively ensure the demise of their own ethos is the great tragedy of the cultural commons.

So where does this leave us? Is it possible to find a happy medium between the isolation and stagnation that results from treating culture as a museum piece on the one hand and the opposing ideal of total "contamination" on the other hand? A compromise that, in the words of Salman Rushdie, "celebrates hybridity, impurity, intermingling," and "rejoices in mongrelization and fears the absolutism of the Pure"?

Noting the proliferation of religious sects in eighteenth-century England, Voltaire was moved to comment that England was a nation of "many faiths but only one sauce." Behind the dry wit was the acute observation that religious diversity, far from being a threat to the public order, was in fact its foundation: "If there were only one religion in England, there would be danger of despotism, if there were only two they would cut each other's throats; but there are thirty, and they live in peace."

For an Enlightenment thinker such as Voltaire, there was no obvious tension between diversity and a harmonious civil society. A liberal polity, Voltaire believed, could tolerate people of any and all beliefs, tastes, and opinions, save those who refused to extend the same toleration to others.

A great deal has changed on the diversity front since Voltaire's day, and it is a bit startling to remember that the multitude of "faiths" that he took notice of were actually all versions of Protestantism. Nowadays, England is a roiling cauldron of competing religiosity: A 2005 survey found that while "Christians" still make up the single largest religious grouping in London (glossing over the internal divisions within that faith) at 58 per cent of the population, there are also around 9 per cent Muslims, 4 per cent Hindus, and 2 per cent each of Jews and Sikhs. This diversity is accepted as a fact of life in the United Kingdom. So much so that Charles, the Prince of Wales, has declared that once he is king he wishes to be known as the "Defender of Faith," to underscore that a vital aspect of his job will be to protect the religious freedoms of all his subjects.

London today is in many ways just a reflection of how diverse

Europe, North America, Oceania, and many other parts of the planet have become. And this diversity is not just religious, as the world rapidly becomes a buzzing confusion of intermixed people of every race, culture, religion, sexual orientation, and lifestyle choice. As social theorist Grant McCracken argues, the world seems to be marching steadily toward Plato's notion of the universe as a plenitude, where every difference proliferates into ever more variety, and "all that can be imagined must be."

This plenitude goes by a number of different names. Call it globalization and people think of free trade and economic dereg- ulation. Refer to it as multiculturalism and you'll find yourself in a discussion over immigration policy and perhaps school curricu- lum. Overall though, the term that best captures all of the various political, economic, religious, and cultural aspects of the ideal is *cosmopolitanism*.

This is not a new word. It was first used, more than likely, by an ancient Greek philosopher (of sorts) named Diogenes of Sinope, better known as Diogenes the Cynic. As mentioned earlier, Diogenes was a bit of a character, and in many ways he could be considered the first philosopher of authenticity. He didn't much like Athenian society and had a great deal of contempt for the city's officials and institutions, whom he considered corrupt. Diogenes believed that the proper life was one of self-sufficiency, devoid of the luxuries and comforts of civilization. And he was no abstract theorist: he was basically a beggar who lived in a tub and spent most of his time wandering around insulting and offending his fellow citizens. He is said to have urinated on people he didn't like, defecated in the theater, and masturbated in the market. (When he was reprimanded for this last act, he replied that it was too bad it wasn't so simple to relieve one's hunger by rubbing one's stomach.)

For all this, he is best known as the father of cosmopolitanism, because when asked where he was from, he answered that he was a cosmopolitan, literally a citizen of the cosmos. This was a pretty radical position to take at a time when virtually everyone believed that every civilized person was a member of a political community, a city or "polis" from which they derived their identity and to which they owed their allegiance. To call yourself a "citizen of the world" was nonsense.

As a political theory, cosmopolitanism lay mostly dormant for centuries, until it was picked up by early liberal thinkers such as Immanuel Kant. Liberals were attracted to cosmopolitanism because it seemed a powerful antidote to a whole tray of political poisons that relied on giving special rights and privileges to a particular race, religion, or class, or that gave special status to those who lived within a geographically defined region. The cosmopolitan rejects both particularism and localism, seeing instead his or her loyalties and obligations as extending to the entire human race. But in addition to aiding in the fight against tyranny, despotism, and other forms of arbitrary rule, there was also a more positive side to the cosmopolitan project. Hearkening back to Diogenes' attitude that it is a philosophy that aids in personal self-development, the liberal cosmopolitans believed that human variety was valuable in itself. As German philosopher Christoph Wieland wrote in 1788, cosmopolitans consider "all the peoples of the earth as so many branches of a single family, and the universe as a state of which they, with innumerable other rational beings, are citizens, promoting together under the general laws of nature the perfection of the whole, while each in his own fashion is busy about his own well-being."

This view was echoed a century or so later by John Stuart Mill in his celebrated essay *On Liberty*. For Mill, diversity matters because it presents people with the raw materials for living, with the range of options they need to make a choice about how to live their lives. His argument was straightforward: just as different plants and animals flourish in different physical environments, different people flourish in different moral and cultural environments. One

man's meat is another man's poison, and unless "there is a corresponding diversity in their modes of life, they neither obtain their
fair share of happiness, nor grow up to the mental, moral, and aesthetic stature of which their nature is capable."

But in recent decades, as the world started to catch up with the
theory – as we took to freely trading with people in other countries, as borders became more permeable, and as formerly stable
and orderly societies became perpetual commotion machines –
people started to have second thoughts. In an ironic twist, as the
spread of liberalism was taking the stuffing out of the racists and
the aristocrats and the religious heresiarchs, local and particular
loyalties started to come back in vogue. For many supposedly
progressive-minded people, cosmopolitanism has become anath-
ema, an enemy of authenticity, and the root cause of countless
social problems.

At the very least, there is a widespread sense that things have
gone too far too fast. The idea seems to be that in a world where
we are loyal to everyone, we are actually loyal to no one, since
having a community or local set of attachments is an essential part
of developing an identity, a sense of who you are and what you
value. Living in a tub, begging, peeing on people – that's fine for
Diogenes. But today's quest for an authentic life requires something more robust in the way of community attachments.

For many people, the single biggest objection to cosmopolitanism
is that it leaves us spiritually and morally bereft. This is part of a
larger complaint about the liberal enterprise, which has as its
central goal the maximizing of the needs of the individual. For a
liberal like Kant, that goal is freedom; for a utilitarian like Mill, it
is happiness or welfare. In both cases though, the fulfillment of
individual needs is paramount, and anything that puts a limit on
freedom or the pursuit of happiness is illegitimate and must be
avoided or removed.

In this view, liberalism is a narcissistic and even nihilistic philosophy, having no conceptual room for values or allegiances that extend beyond the whims or desires of the self. One of the starkest accounts of this view is found in the novel *Les Particules élémentaires*, which has been variously translated into English as *The Elementary Particles* and *Atomised*, by French writer Michel Houellebecq.

The story is a dystopian, quasi-science fiction narrative about two half-brothers, Bruno and Michel. Michel is a molecular biologist, while Bruno is a high school teacher who has turned into a sex addict. Both are sad, screwed-up loners, largely thanks to the pathetic and mostly nonexistent upbringing by their hippie mother. Much of the book consists of Bruno and Michel sitting around the supper table with their girlfriends talking about how lousy their lives are, but it is set against the backdrop of the emergence of cloning as a replacement for human sexual reproduction (which is itself a consequence of Michel's pioneering work). With basic human connection no longer needed even to simply reproduce the species, humans are revealed as nothing more than autonomous monads, bouncing around the universe interacting with one another but never really connecting.

It was a bit of a scandal when the book was published in 1998, largely due to its rather graphic sexual descriptions, though it didn't help (or hurt – the book sold hundreds of thousands of copies) that the author has a roguish personality that matched the book's content. Regardless, *Les Particules élémentaires* was hailed by many reviewers as a classic account of the self-centered nihilism at the core of liberal society.

What supposedly makes the particularism of community so valuable is that it shows us a path away from the liberal obsession with the self toward the love of something greater. By being members of a community of family and friends, of people who share our history, religion, and values, we move away from the short-term and insistent demands of the self and become attuned to the needs of others. Against the liberal, the communitarian says "it's not all about you." It is about recognizing that there are goods that transcend our individual wants, that command our allegiance, and might require our sacrifice.

It's a familiar charge that liberal cosmopolitanism is a short hop from nihilism, but a misguided one for all that. Diogenes may have been a misanthrope, but for the classic liberal thinkers the appeal of cosmopolitanism was never that it permitted us to be self-centered and narcissistic. Just the opposite! Liberals liked the way it made us aware of new values, new forms of living, and helped breed the virtues of tolerance, respect, and mutual regard. As Kwame Anthony Appiah puts it, cosmopolitans think human variety matters because people are entitled to sample the options they need to shape their lives *in partnership with others*.

One of the difficulties in sorting out the dispute is that the cosmopolitans and the communitarians are in many ways talking at cross purposes. The cosmopolitans are talking about the importance of certain liberal *principles*, while the communitarians are concerned about the effects of that liberalism on *values*. The principles/values distinction is one that we don't usually make in our everyday language, and it is not uncommon for the terms to be treated as virtual synonyms (a man of principle is often respected for having strong values, for example). Academics, however, find it useful to distinguish between the two because each focuses on a different aspect of our moral lives.

When we talk about principles, we are referring to the general rules that govern our sense of right, such as the constitutional statements of due process, equality, and freedoms of religion, speech, and association that support a liberal society. Values, on the other hand, refer to our sense of what is good to do or believe. Values are what give our lives meaning or purpose.

To make the distinction a bit clearer, the liberal principle of free speech allows you to express yourself as you choose, while your values will determine the manner in which you choose to express yourself. When Pierre Elliott Trudeau said that "the state has no business in the nation's bedrooms," he was expressing the principle that sexual behavior is no business of the government. What Trudeau had no official position on were the sexual values that people might have – abstinence, monogamy, or swinging poly-sexual foursomes. Yet as many people will admit, fulfilling our diverse sexual needs is part of what gives life meaning. What Appiah is doing is affirming the principle that people ought to be allowed to sample as much of what humanity has to offer in order to discover just what values are worth holding. Diversity is useful because it provides us with new sources of potential value.

As far as the communitarian is concerned, the problem is that some lives can only be pursued in the context of shared values, as a member of a community of like-minded people. It is easy to see how this is true for some groups, such as the Amish, who live in an agrarian society based on the separation from the world and the collective rejection of certain forms of technology. It does not take much imagination to see how an Amish community would quickly fall apart if some members suddenly started using industrial com-bines or if a handful of families allowed their kids to have cell phones and high-speed Internet connections.

Very few of us are as isolationist and guarded as the Amish, yet many still see the appeal of living in a community surrounded by people who share their values. To a large extent, this is driven by the idea that every community of any given size is ultimately held together by shared values. So what keeps your neighborhood from descending into a war zone is the fact that it is made up of people who have a shared understanding of the good life. Security, comfort, trust, friendship – these can only be preserved in a com-munity where everyone has a common set of values.

And so, contrary to Voltaire's belief that diversity is a source of social stability, there is the work of American sociologist Robert

Putnam, who for the past decade has been engaged in a substantial study on the relationship between social cohesion and ethnic diversity. According to Putnam, in the face of large-scale immigration and increased diversity, people tend to "hunker down" and "act like turtles," while the bonds of community are stretched to the breaking point as people become less open and more distrustful of one another. Worse, Putnam found that it is not just that diverse communities tend to fragment into a checkerboard of homogeneous but tight-knit clans. Rather, trust breaks down across the board. As Putnam puts it, the problem is not that in diverse communities we don't trust people who don't look like us, but that we don't even trust people who *do* look like us. And this basic lack of trust manifests itself in all sorts of ways, including lower confidence in local government, lower voter turnout, less cooperation and participation in community projects (such as neighborhood watch programs, or public gardens), less charitable giving, and more time spent watching television.

When the bleak implications of his research were first widely reported in 2006, Putnam found himself with an unfamiliar group of friends: racists, anti-immigration activists, and right-wing nationalists who thought that they'd been given a hefty new club with which to beat the soft skulls of multiculturalists. For a self-styled progressive such as Putnam, this reaction was agonizing, since he does not believe that America, Britain, or any other country needs to do a major rethink of its open policies toward immigration and its support for diverse, integrated neighborhoods. He argues that if increased diversity leads to the corrosion of community in the short term, our long-term project must be to construct a broader and more inclusive civic identity. As he points out, America did it once – in the first half of the twentieth century, when it assimilated millions of Irish, Italians, Poles, Germans, and Jews – and there is no reason why it cannot do so again.

For many critics of diversity, however, this is pure wishful thinking.

In the early hours of March 13, 1964, a twenty-eight-year-old woman from Queens, New York, called Kitty Genovese arrived home after a long shift at the bar she managed. As she made the short walk from her car to her apartment building, she was approached by a business machine operator named Winston Moseley, who chased her down and stabbed her twice in the back. Moseley ran off after Genovese screamed for help, and she slowly dragged herself toward her building. But ten minutes later Moseley returned, and after a brief search, he found Genovese lying semi-conscious in a hallway. He stabbed her a few more times, sexually assaulted her, robbed her, and left her for dead. Police and ambulance personnel arrived a few minutes later, but Genovese died before reaching the hospital.

The murder of Kitty Genovese was a sensation, almost immediately held up as an example of the anonymity, alienation, and complete lack of mutual regard in urban America. This inflammatory view of her killing was formed by an article published in *The New York Times* under the headline "THIRTY-EIGHT WHO SAW MURDER DIDN'T CALL THE POLICE," and which began, "For more than half an hour thirty-eight respectable, law-abiding citizens in Queens watched a killer stalk and stab a woman in three separate attacks in Kew Gardens." The story also quoted a neighbor as saying that he didn't call police because he "didn't want to get involved." The writer Harlan Ellison later added his own spin to the story, claiming that one of Genovese's neighbors even turned up his radio in order to drown out her screams.

Neither the *Times* nor Ellison gave an accurate account of what happened that night. The total number of Kew Gardens residents who heard the attack and Genovese's screams was probably around a dozen. Not one of them heard or saw the entire grisly affair from start to finish, and many who did hear Genovese screaming thought they were eavesdropping on a domestic dispute, not a murder.

Nevertheless, the killing of Kitty Genovese launched an entirely new line of research into the diffusion of responsibility in large communities, which came to be called the Bystander Effect. The gist of it is that, paradoxically, the more people who witness a crime or see someone in need, the less likely it becomes that anyone will stop to help. The case continues to be discussed in introductory texts in sociology, while in the popular imagination it has become an urban parable warning of the callousness and alienation of urban life.

The sheer brutality of the killing does not explain the ongoing interest in the story. There is something about the case that seems to strike the resonant frequency of city living. Think of the number of times you've heard a car alarm going off and not bothered to investigate, stood in a crowd and watched a building burn, or seen a car stopped by the side of the highway with its hood up, the driver poking helplessly around the engine, and simply kept driving. For the most part this isn't indifference or callousness, it is simple rationality. We assume that with so many other people around, someone else must have already gone to help or called 911. In many ways, the Bystander Effect is a type of collective action problem, where private, individual rational calculations lead to a collectively unfortunate set of outcomes.

This tells us something interesting about the deep nature of the bonds of trust and sociability that form the core of what most people mean when they talk about "community," and why it seems to be absent in urban settings, especially ones high in cultural or racial diversity. It turns out that the problems with diversity, as exposed by Robert Putnam's research, do not stem from racism, classism, prejudice, or any xenophobic refusal to associate with the "other." Instead, they appear to be based on simple rational calculation.

Let's start with a straightforward example: the practice, which is now hugely degraded, of holding a door for someone going in or out of a public building. This is usually thought of as a matter of what used to be called "common courtesy": I don't know you, I'll never see you again, but I take a moment out of my busy day to hold the door for you as you enter behind me. But there is more to our usual understanding of common courtesy than selflessness or altruism. Holding the door for someone is not seen as praise-worthy, because you're *supposed* to hold the door for people, and the failure to observe these sorts of minor social graces is seen as a moral failing. Things like holding the elevator when someone is running for it, making eye contact with the cashier at the super-market, cleaning off the treadmill at the gym after you've spent an hour sweating all over it – these are the other-regarding acts that buffer our daily interactions as we knock around the city like billiard balls.

It is the degradation of these norms, the way they have come to be honored almost entirely in the breach, that fuels the sense that our public life has become more impersonal and alienating. We romanticize life in small towns, where we imagine that the cashier always smiles as she hands you your change, the bartender always knows your drink, and no one ever has to run to catch an elevator. To understand why this is the case, it helps to remember that our behavior is shaped by the fact that nothing we do happens in iso-lation. It is shaped by how other people have treated us before, and how we expect other people to treat us in the future. All of these courtesies have an element of reciprocity to them. Someone holds the door for you, and then you hold it for someone else. Someone cleans off the elliptical machine at the gym before you get on, and you clean it for the next person when you get off.

Now, if these alternations are frequent, and occur between the same pairs of individuals over an extended period of time, the prac-tice evolves to the point where it becomes something like a private exchange in a barter economy. When you hold the door for me, you are in a sense "paying me back" for holding it for you yesterday. If you fail to do so, it is very easy for me to punish you by simply refusing to hold the door for you tomorrow, though that would make for very uncomfortable future interactions. But because we see each other every day, there is no reason to defect from the door-holding economy – each of us can be quite certain that we will never be stiffed and left in a door-holding deficit.

It is easy to see how this same pattern of explanation can be applied to many of the other common courtesies. It also explains why your local corner store might offer you credit or let you write a check, or why your regular watering hole will let you run a tab. The general point is that in a small, close-knit society, or in any other relationship that offers regular interaction, compliance with basic norms of social courtesy can be rather easy to enforce.

Common courtesy can also be sustained in a larger and more urbanized society, as long as it remains culturally, ethnically, or racially homogeneous. In part, this is because the shared background social norms and expectations are already in place. Social norms vary from place to place and from one culture to another, and learning how things are done "around here" is one of the toughest things about moving to a new part of the country or traveling to a foreign land.

But there are other reasons why social norms of trust and sociability can be sustained in large but homogeneous societies. For one thing, such societies are more likely to be held together by overlapping institutional relationships, work experiences, and family ties. In a society still governed by relationships forged in common workplaces, schools, churches, and civic bodies, or through extended families and intermarriage, the glue of trust will often retain its strength over large distances and time frames.

But think of what happens when the society in question becomes larger and more ethnoculturally diverse. First, there are going to be fewer shared norms, especially with respect to the way men and women are supposed to interact. Do you hold the door for a woman or does the man always precede the woman when entering a room? Do men and woman shake hands, kiss on one cheek, both cheeks, or not touch at all? But more important is the way the tight loop of reciprocity in our social interactions gets stretched and twisted when we start to see one another less frequently. There is no longer the sense of it being a barter-based system, where one courtesy is promptly repaid with another. As the number of people involved in this social economy increases, it becomes harder to keep track of who owes what to whom.

And so as the size of the courtesy market grows, simple acts like door-holding switch from a system based on private mutual exchange to something more like a system based on a public good. There emerges a "door-holding commons," with individuals contributing their door-holding expertise for the benefit of whoever happens to be walking in behind them. And like many public goods, it is easily undermined by free-riders, people who happily benefit without contributing. The tragedy of the commons quickly shows its ugly face, as people start barging through entranceways, letting the door swing shut in the face of the anonymous folks behind them.

Again, it is easy to see how the breakdown of the trust economy can lead to a more general decline in all forms of reciprocity, sociability, and mutual aid. Why stop to help someone who is trying to change a flat by a busy highway in the pouring rain when you'll probably never see that person again? Besides, what guarantee do you have that someone would do the same for you? Heck, maybe it is a ruse and they want to rob you. Sure, everyone would like to live in a more trusting, secure, and sociable environment, but no one wants to be played for a sucker either. In this rather bleak view of things, only strongly internalized social values (such as those described by the parable of the Good Samaritan) along with dying vestiges of natural sympathy, keep the streets from degenerating into a thoroughly Hobbesian war of all against all.

The demise of the trust economy does not mean that people stop wanting social goods. We still want to go to a bar where someone knows our name, we would like a smile when we purchase our coffee. The need for someone to call the police if you are in trouble does not go away, and people still need help changing flat tires. But when we can no longer get these things through the informal barter mechanism of the trust economy – that is to say, from our community – we turn to more formal instruments.

One of the most common of these is the market. When we desire goods that are no longer provided through social relationships or the bonds of community, many people simply buy them. One of the clearest examples of this is the need for home security. When you can't trust that your neighbors won't rob you while you are away or asleep, then you'll need an alarm system, if not someone to patrol the grounds. If you live in an apartment building or condo, you can hire security personnel to man the front desk and check who is coming and going. If you are rich enough, you can move into a gated community, which is the modern equivalent of a walled city.

But the increasing shift to a reliance on the market is everywhere these days. Many of us no longer live in the same city as our immediate family, so we hire babysitters or nannies to take care of our kids. If you are single and having trouble meeting someone, you can use an online dating service. If you are alone and haven't any friends, you can go to the local bar, where, for the price of a couple of beers, you can chat the bartender's ear off all night. The extreme end of this spectrum is prostitution, where what is supposedly the ultimate act of intimacy is turned into a quick, anonymous commercial transaction.

Given this apparently bleak picture of urban life, it is no wonder so many people are fleeing the cities, looking for a reasonable facsimile of community far from the madding crowd. Yet what began as a well-intentioned flight from polluted and congested cities has come to be seen, in the eyes of many critics, as an evil unto itself. It's called suburbia.

The heart of the case against the suburbs, the principle from which all else follows, is that they are inauthentic. Suburbia, goes the critique, is fake, artificial, unreal, and ersatz, offering a mere simulacrum of real living to people who are either too dim or too brainwashed by advertising to know better.

It wasn't meant to be like this. Until the mid-eighteenth century, there were people who lived in the city and people who lived in the country. Those who lived in the country did so for pretty much one reason: agriculture. Folk clustered in the city for a variety of reasons, mostly having to do with the need for proximity and security. Politics and commerce need to be conducted face to face, and the absence of rapid transportation meant that people engaged in these functions needed to live close to one another. At the same time, the desire for security meant that people had to live inside the city walls. As a result, the features of urban life that many people today see as ideal – high density, mixed-use neighborhoods – were in fact a necessary evil of urban life.

The earliest moves toward what we would today call "sprawl" were the villas and country manors of the wealthy and the aristocracy, who could afford the privilege of escaping the noise, stench, and congestion of the city. Between the First and Second World Wars, the ability to escape the city gradually morphed into a middle-class privilege. A major catalyst in many North American cities was the streetcar, which for the first time made it possible for a large number of people to get from the downtown to the outskirts and back, in a reasonable period of time and at a reasonable price. Housing developments sprung up along the streetcar lines as more and more people abandoned the commotion of the city for quieter developments of row houses, bungalows, and semidetached homes. This process was well underway by the 1920s, but it was interrupted by the depression and then by the Second World War. Once the war was over though, the decanting of the city to the suburbs began once again, encouraged this time by the twin forces of big business and big government.

As it is usually told, the story of the postwar outward migration has a conspiratorial edge to it. After the war, America had a huge industrial and manufacturing surplus capacity that could not be soaked up by the devastated economies of Europe. So the Detroit car makers teamed up with the oil companies and the housing industry to convince young couples that their future happiness could not be found in the dirt and grime of the city but in the pastoral gentility of the country. To get there, though, they were going to have to drive. To cement their role as the midwives of the American dream, General Motors, Standard Oil, and a few other companies bought the streetcar lines in a number of cities and tore up the tracks, while the federal government instituted a massive road and highway construction plan that connected the increasingly hollow city cores with the booming suburbs.

Thus did the American way of life come to be identified with the automotive way of life, and these pioneering families found themselves living in cookie-cutter developments that offered neither the convenience and community of the city nor the privacy and charm of the country. Instead, they were stranded in a no man's land, a vacant and sterile world from which the only means of escape was the automobile.

This has become the received view of suburbia. It holds that the suburban lifestyle is more than just aesthetically unappealing; it is also responsible for a great deal of our modern woes, including social alienation, depression, obesity, consumerism, the demise of the family farm, smog, power blackouts, global warming, and the global war on terror. We have to keep in mind though, that despite it being so familiar, this attack on suburbia is only a theory, incorporating a combination of historical narrative, social criticism, and economic conspiracy theory. The question, though, is whether any of it is true.

The conviction that suburbia is not, in a fundamental sense, a real place flows through the entire postwar critique. It emerged as an offshoot of the 1950s critique of mass society, and it has been a leitmotif of the antisprawl literature ever since. Perhaps the most influential (if strident) voice was that of Lewis Mumford, the mid-century social critic who was one of the first thinkers to seriously examine the connections among technology, culture, and social institutions. One of his most enduring works is the 1961 book *The City in History*, which serves up a deeply negative and pessimistic account of contemporary urban development.

In particular, Mumford was fond of denouncing the homogeneity and formlessness of suburban developments in language that was as unequivocal as it was hyperbolic. In one frequently quoted passage, he writes that suburbs consisted of

> a multitude of uniform, unidentifiable houses, lined up inflexibly, at uniform distances, on uniform roads, in a treeless communal waste, inhabited by people of the same class, the same income, the same age group, witnessing the same television performances, eating the same tasteless pre-fabricated foods, from the same freezers, conforming in every outward and inward respect to a common mold.

For Mumford, these aesthetic deficiencies were just the outward face of a deeper problem, which was the lack of a proper "organic" relationship between people and their lived environment. With their emphasis on the car and lack of genuine mixed-use neighborhoods and proper communities, the suburbs' uniformity led inevitably to repression and alienation.

Much of the vocabulary we instinctively reach for when talking about the aesthetics of sprawl is thoroughly Mumfordian in both tone and content. Meanwhile, the notion that the aesthetic uniformity of suburbia breeds psychological repression and social conformity is the Old Faithful of the world of cinema and letters, a reliable trope that gives a veneer of apparent depth to films (*The Ice Storm, Revolutionary Road*), television shows (*Desperate Housewives*), novels (*The Corrections*), and music (the Green Day concept album *American Idiot*).

But if there is one figure tending the flame of Mumford's critique, it is novelist and journalist James Howard Kunstler, whose 1994 book *The Geography of Nowhere* is one of the more popular restatements of the critique of suburbia as an alienating and inauthentic form of living. Kunstler isn't one for nuance, and his description of suburbia as "the greatest misallocation of resources in the history of the world" seems a bit over the top (the Nazi gas chambers and the Soviet gulag come to mind as counterexamples) until you realize that he's simply trying to outdo his hero Mumford's prediction that the inevitable consequence of unchecked sprawl would be nuclear holocaust.

In *The Geography of Nowhere*, Kunstler argues that the two main problems with suburbia are the extreme separation of land use and the large distances between places that result. Because the suburbs segment our living space into highly defined zones dedicated to either residential, commercial, industrial, or business uses, people end up having to drive all over the place as they go to work, do their shopping, and go home to sleep. This is a waste of time, it is expensive, and it causes pollution. But far and away the most significant impact, says Kunstler, is "the sacrifice of the sense of place." The feeling of an organic, authentic connection to the world is gone, he says, and there are no more "sacred places, places of casual public assembly, and places of repose."

Like Mumford before him, Kunstler believes that what people really want is a smaller, more human scale and local form of living.

The ongoing nostalgia of Americans for the dimly remembered small-town life of the first half of the twentieth century is nothing more than a longing for a form of true community, an escape from the living nightmare of mass society:

> When Americans, depressed by the scary places where they work and dwell, contemplate some antidote, they often conjure up the image of the American small town. However muddled and generalized the image is, it exerts a powerful allure. For the idea of the small town represents a whole menu of human values that the gigantism of corporate enterprise has either obliterated or mocked: an agreeable scale of human enterprise, tranquility, public safety, proximity of neighbors and markets, nearness to authentic countryside, and permanence.

There are a number of problems with this argument, the most salient of which is that the distinction between what is "real" and what is "fake," or what is authentic and what is inauthentic, tells you a lot more about the person making the judgment than it does about what is being judged. As architect Witold Rybczynski puts it, the claim that the suburbs are not "real" makes sense only if one assumes that a real city has cathedrals and plazas instead of parking garages and fast-food franchises, sidewalk cafés not shopping malls, live theatres instead of cineplexes. The suburbs are fake only if you assume that what is "real" is beautiful and high brow and what is fake is unfinished and dedicated to mass taste.

For the past decade or so, a small group of academics and urban planners has been engaged in a fundamental rethink of the biases and prejudices built into the received view of the relationship between city, suburbs, and exurban sprawl. One of these is Robert Bruegmann, whose book *Sprawl: A Compact History* deliberately sails directly into the prevailing wind of popular opinion. Unlike most people who study the causes and consequences of suburban sprawl, Bruegmann has no difficulty attributing it to the choices and desires of the people who actually move there, and he doesn't condescend to suggest that they've been bamboozled by advertising or manipulated by the oil companies.

According to Bruegmann, the received view of suburbia completely misunderstands what people are after when they leave the downtown and head for the outer boroughs. Contrary to Kunstler's caricature, they don't imagine that they will live like country squires, nor do they delude themselves into thinking that a development named "River Heights" or "Creekside" will be above the river or beside the creek. Instead, they move to the suburbs in search of three prosaic goods: *privacy*, *mobility*, and *choice*.

By privacy, he means the ability to manage your living space and the people who have access to it. Mobility is just the ability to get around, to go where you want, when you want, through a means that is under your control. Finally, the suburbs allow people to choose from a variety of lifestyle options with respect to where they live, work, and engage in recreational activity. What is really at issue here is the question of control, and what is important to note is that for a long time, only the wealthy had enough control over their environment to guarantee their privacy, mobility, and choice. For example, the detached home now offers the same degree of privacy as a high-rise co-op, while the automobile provides the sort of private transportation that was once available only to those who could afford a private carriage.

This all seems pretty obvious, in retrospect. So what explains the surprising persistence of antisuburb sentiment and its blatant contempt for the taste and choices of the middle-brow masses? Robert Bruegmann suggests that the virtue of the democratization of privacy, mobility, and choice is discounted by members of the cultural elite. The elites are not interested in hearing about the benefits of increased choice and control for the public at large "because they believe that ordinary citizens, given a choice, will usually make the wrong one."

It is hard to avoid concluding that virtually the entire case against the suburbs is little more than lifestyle snobbery disguised as a quest for authenticity. Or more accurately, it appears to be a classic instance of the desire for authenticity revealing itself as a thinly veiled form of contempt for middle class tastes and preferences. Lewis Mumford's line about anonymous clans of stupefied drones watching the same television shows and eating the same prefab food drips with high-brow condescension, but he probably didn't care. Mumford never really tried to hide his elitist belief that suburbanites were simply too dim to realize how they had been had, too illiterate to read the fine print on the bill of goods they had been sold. Parroting as always his master's voice, James Howard Kunstler blames the mass media for having more or less brainwashed people into living in what he calls a "consensual trance." Professing himself mystified about why people can't see the problems with suburbia, he suggests that maybe it is just too difficult to get the attention of Americans, to distract them from "recreational shopping at the mall, and Jennifer Lopez, and playing computer games, and NFL football, and NASCAR."

Yet the people who move to the suburbs aren't nearly as stupid or careless or brainwashed as the urbanites seem to think. They know they're going to get a lawn, a garage, and a backyard. They know they will be miles from a store or cafe, and that they'll have to drive everywhere. Most people move to the suburbs with eyes wide open, fully aware of the tradeoffs they are making. They are looking not for some pastoral idyll, but for more privacy, space, quiet, and parking. People have been trying to flee the city for as long as there have been cities, and the only thing that distinguishes sprawl today and the sprawl of yesteryear is the social status of those who move there. Until quite recently, cities were dirty and dangerous, so the rich took their advantage and escaped to the countryside. Today, our cities are safe and clean, increasingly populated by the hip, the young, and the childless, while the suburbs are for those unwise enough to have children or vulgar enough to desire a driveway.

As Bruegmann puts it, sprawl always seems to be someone else's fault. It is "where other people live, the result of bad choices and poor judgment by other people." Sprawl is like bad taste. No one admits to having it, yet somehow there seems to be a tremendous amount of it around.

The notion that the decision to live in the suburbs involves trading off one set of lifestyle preferences against another is in many ways just a microcosm of the modern dilemma. In the face of the community-destroying forces of globalization, the question we should be asking ourselves is not "What can we do about it?" but rather, "How should we think about it?" As with modernity as a whole, with respect to cultural change and cosmopolitan contamination, we need to think in terms of tradeoffs: losses to be sure but also benefits.

The globalization of trade, and especially of the trade in cultural goods and services, tends to have the following effects. First, by drawing previously isolated, autarkic, or otherwise independent communities into the global economy, it tends to lead to increased diversity *within* cultures at the expense of diversity *between* cultures. You can get sushi in Montreal and poutine in Tokyo, but the result is that serious "otherness" becomes harder to find around the planet even as our own countries – and our immediate surroundings – become increasingly diverse.

Second, as the number of distinct cultural identities shrinks, so does the set of possible comprehensive worldviews or ideologies. The last major alternative to liberal democracy and market capitalism was Marxism, and it collapsed in 1989, economically exhausted and morally bankrupt. Notwithstanding a few notable exceptions, such as Cuba and China (and perhaps some backsliding in Russia), liberalism is advancing steadily across the globe. But what liberalism takes it also gives back, through an explosion of diversity within Western civilization as a whole.

This leads to a third and possibly worrisome trend, which is a steady shift in the way we consume culture, from *focused and deep* to *wide and shallow*.

The advantage to living in a homogeneous society, where there is a widely shared cultural identity, is that it permits a deep knowledge and critical appreciation of what the culture has to offer. For example, it is thanks to the Italians and their deep love of opera that opera remains a living artistic tradition. Lovers of good food owe a debt of gratitude to the French, whose impossibly rigorous demands for fine dining have set standards that the chefs around the world are obliged to pay attention to. It is not hard to multiply these sorts of examples – Americans and jazz, Russians and literature, Japanese and Zen – we are able to appreciate these as living artistic traditions because of the way they have been sustained by domestic consumers and critics who understand them in their fullness and depth.

But in an age of cultural plenitude, the temptation is to spread our attention as widely as possible. The world is increasingly seen as a buffet of tastes and sounds and ideas, and we gorge ourselves without pause or consideration. We treat centuries-old traditions with the same respect and attention as the latest offering from the latest disposable pop starlet; it is all grist for the rip/mix/burn culture of the worldbeat cosmopolitan. We become like cultural cattle, grazing from one pasture to the next without ever getting to know or understand the richness of the soil.

Put these three interlocking trends together and you arrive at a version of the now-familiar complaint about globalization, that it leads to a homogenized, shallow world dominated by disposable culture and consumerism. The reality is somewhat more nuanced, and it must be stressed that these are mere trends, not certainties or absolutes. Still, the complaint is not without merit. What it lacks, though, is a proper appreciation for the way the protection of a community's ethos or identity can only come at the cost of severe restrictions on the choices and freedoms of the individuals in the community. To see this in action, just look at Bhutan.

If you check it out in an atlas, there is nothing terribly remarkable about the Kingdom of Bhutan. About half the size of Indiana, it is a mountainous country of fewer than 700,000 people, most of whom are Buddhist, nestled between China and India. Its economy is largely based on subsistence agriculture with some forestry, with a per capita GDP of around $5,200 per year and a literacy rate of only 47 per cent. All told, the country ranks a rather woeful 134 out of 177 countries on the UN's Human Development Index, just below Congo and Namibia but ahead of its neighbor India.

But this doesn't faze the Bhutanese. Since 1972, the country has been committed to the notion of "Gross National Happiness," or GNH. The idea behind GNH is simply that there is more to life than money, and that development needs to be evaluated in a more organic fashion, incorporating material and spiritual growth, protection of the environment, and orderly government. Although the Bhutanese government has helped sponsor a number of conferences on the subject, GNH remains a poorly defined concept, and by some counts there are as many as twenty different measures floating around. But in what was widely seen as a vindication of its efforts, Bhutan ranked eighth overall in a 2007 study of subjective well-being (a.k.a. happiness) in 178 countries. It was the only low-GDP country to come in the top twenty, and it ranked well above other countries at comparable levels of economic development.

Regardless, the actual measure of happiness is probably less important than what it represents, which is the ongoing determination by the country's rulers to remain aloof from the modern world and to preserve Bhutan's distinctive ethos, its cultural and religious identity. To that end, the government has long maintained strict controls over industrialization, technology, trade, and communications. The country has no railway, and television and the Internet were illegal there until 1999.

One of the most interesting restrictions is on tourism. The Bhutan Department of Tourism has complete discretion over who gets into the country, and while a visa only costs around $20, visitors have to prebook a tour package for a minimum of five days at $250 per day. As a result, tourism is now the third biggest industry in the country, and it has led to the well-founded suspicion that GNH is as much a marketing slogan to attract rich foreigners as it is a statement of national purpose. A visit to the "last Shangri-la" on Earth definitely provides an alternative perspective on what is important in life, but it literally is not for everyone. In effect, the country has turned itself into one giant eco-Buddhist luxury resort for well-heeled authenticity-hunters looking to avoid the teeming masses of sweaty, bearded backpackers in more colonized parts of Asia. Or, as the *Lonely Planet* guide puts it, Bhutan is "Nepal for the jet set."

There are a number of reasons why Bhutan has been able to remain so isolated for so long, one of the most important of which is simple geography. The place is landlocked, with very high mountains to the north and west, and the eastern frontier borders the sparse and relatively unknown Indian state of Arunachal Pradesh. When you combine geographic isolation with widespread poverty and illiteracy, high rates of religiosity, and severe restrictions on communication from the outside world, you are left with a population that is ignorant, docile, and therefore easy to manipulate.

All of that is changing. Over the past decade, Bhutan has changed from an absolute monarchy to a constitutional monarchy run on broadly democratic principles. The Bhutanese, especially the youth, are getting their first taste of freedom and a better picture of what life is like elsewhere, and this is causing dissatisfaction with the status quo. Bhutan is trying to engage modernity on its own terms, but eventually, the country's youth are going to start wearing Nikes, talking on cell phones, and listening to hip-hop. Bhutan may succeed for a while in maintaining important elements of its ethos, but ultimately modernity and economic development will take their destructive toll. And eventually the tourists will stop coming.

What all of this underscores is the deep contradiction between liberal cosmopolitanism and the communitarian desire to preserve distinct cultural identities. For the communitarian, the effect of opening up on a country such as Bhutan is a strike against cosmopolitanism, since any system that so relentlessly works to undermine a people's cultural heritage must be dangerous and should be resisted. But we can just as easily throw the ball back and shift the burden of argument. Why should the community be our prime unit of concern? Does the community exist to serve the people or vice versa?

Any community whose survival rests on keeping its members ignorant, poor, isolated, and politically disenfranchised should not be worthy of our respect or our efforts. It may warm the hearts of authenticity seekers and left-wing academics to see Havana frozen in 1946, just as we may wish that Bhutan remain a charming kingdom of agrarian Buddhists, toiling away in their rice fields in the foothills of the Himalayas. But we need to recognize the people of Bhutan and Cuba for what they are: victims of the hoax of authenticity and pawns of people who would put their culture in a museum as a symbol of resistance to the modern world.

But even this concedes too much to the opponents of cosmopolitanism, because it considers only what is being lost. Missing from this argument is any sense of what is gained in the way of political freedoms, increased material wealth, and greater intellectual and creative opportunities. More to the point, it ignores the way that cosmopolitanism represents not only a material advance, but also a form of moral progress.

The moral core of cosmopolitanism is a universalism rooted in an expansive toleration. This is the belief that everyone matters and that the scope of our obligation to one another is not limited by skin color, language, ethnicity, religion, or any other morally irrelevant fact. Along with this goes a healthy respect for pluralism – the acceptance that there are many ways of finding happiness in this life, many paths to the good life, and that these may not be compatible or reconcilable. Still, the recognition of this deep diversity instills in the cosmopolitan a sense of fallibility, and the understanding that we may be wrong about our judgment of what is valuable and what is worthwhile, or that our knowledge may be only partial or provisional.

The cosmopolitan ethos is a mansion of many rooms, haphazardly fashioned out of what Kant called "the crooked timber of humanity." But all are welcome, as long as they are willing to abide by the terms of use. If it turns out that cosmopolitanism is not, in the end, compatible with someone's preferred account of what constitutes an authentic culture, then Kwame Anthony Appiah has it right. So much the worse for authentic culture.

EIGHT

THE END OF HISTORY

ONE DAY, SAMUEL JOHNSON AND HIS FRIEND AND BIOGRAPHER James Boswell were debating whether Boswell's fondness for London would diminish if he were to move there from his home in Scotland. Boswell worried that the excitement and energy he felt on his occasional visits to the metropolis would fade with familiarity, a suggestion that Johnson famously dismissed: "Why, Sir, you find no man, at all intellectual, who is willing to leave London. No, Sir, when a man is tired of London, he is tired of life; for there is in London all that life can afford."

This conversation took place in 1777, and if anything, the last clause in that statement is truer now than it was in Johnson's day. London remains one of the world's great cities, a place where virtually all there is on offer in the way of excitement, entertainment, and stimulation can be had. No matter what your tastes and proclivities, if you desire it, the chances are London can provide it.

Johnson's "tired of London, tired of life" line has been endlessly quoted and paraphrased over the years, usually to describe another place, event, or activity that is so utterly awesome and fulfilling that to tire of it would be to tire of existence itself. One of the most insightful repurposings is found in Douglas Adams's book *The Restaurant at the End of the Universe*, in the description of the planet Ursa Minor Beta (home of Megadodo Publications, the original publishers of the *Hitchhiker's Guide to the Galaxy* guidebook). According to the guidebook's description of the planet: "Although it is excruciatingly rich, horrifyingly sunny and more full of wonderfully exciting people than a pomegranate is of pips, it can hardly be insignificant that when a recent edition of Playbeing Magazine headlined an article with the words 'When you are tired of Ursa Minor Beta you are tired of life,' the suicide rate there quadrupled overnight."

It's a toss-off line, one of dozens of little in-jokes sprinkled throughout the *Hitchhiker's Guide* series. But these are the jokes where Adams frequently buries his deepest insights, in this case, about the disturbing relationship between the ecstatic pleasures of the modern world and a crushing existential boredom that underscores the essential meaninglessness of life.

There is a lot about everyday life that is boring. Standing in line at the store or the bank, waiting for the bus or the train, sitting at stop lights – it's all dull as doing the dishes. Work is repetitious, unchallenging, and unoriginal. Car trips are boring, as is staying at home. Our friends bore us with the same old stories, our lover bores us with the same old moves. Society is formed of two mighty tribes, Byron wrote, the Bores and the Bored, though the more plausible truth is that each of us is a paid-up member of both clans.

There are plenty of ways to head off the boredom that creeps into the interstices of everyday life. Carry a book with you, join a gym, cultivate a hobby, plan a vacation, and otherwise find ways of keeping your mind occupied and your attention focused. Thanks to consumer electronics, we can now fill even the most transient empty moments with music, chat, videos, and news, to the point where many people wouldn't dream of leaving the house now without an iPod or cell phone. Even Barack Obama – a man who you would think had plenty to occupy his thoughts – kicked up a huge fuss because they tried to take away his BlackBerry when he became president.

Philosopher Martin Heidegger called this constant search for novelty and stimulation the sickness of the modern age, claiming that it was a reflection of an existential fear of being bored. He was probably right. In a commencement address he gave at Dartmouth College in the mid-1990s, poet Joseph Brodsky warned the graduands that if they thought university could be boring, even the most monotonous lecture or textbook was nothing compared to "the psychological Sahara that starts right in your bedroom and spurns the horizon." Like Heidegger, Brodsky believed that the flight from boredom is what spurs the constant search for invention and originality, though he saw it as a little more than a long trip on a hamster wheel:

> Basically, there is nothing wrong with turning life into the constant quest for alternatives, into leapfrogging jobs, spouses, and surroundings, provided that you can afford the alimony and jumbled memories. This predicament, after all, has been sufficiently glamorized onscreen and in Romantic poetry. The rub, however, is that before long this quest turns into a full-time occupation, with your need for an alternative coming to match a drug addict's daily fix.

Brodsky suggests that we simply embrace our boredom and revel in what it reveals about time, existence, and meaning. To wit: time is infinite, existence is fleeting, and our lives are meaningless.

This is very stiff medicine indeed, and while it may charge the constitution of a Nobel Prize–winning poet, not everyone is psychologically equipped to deal with insignificance, futility, and meaninglessness. How many of us can stare into the abyss and not come out of it psychologically damaged? The brutal implication of Douglas Adams's "joke" about life on Ursa Minor Beta is that if this is the best life has to offer, then maybe life really isn't worth living at all. Excitement, entertainment, and constant stimulation are the empty calories of the spirit, and eventually we have to come to grips with the final consequences of the disenchantment of the world. What's it all about? Nothing, comes the cold-blooded answer. What, then, is worth doing? That's the question modernity poses. And coming up with a credible answer is our foremost task here, at what has come to be known as the End of History.

It was 1989, and Mikhail Gorbachev was still in power, the Soviet Union still functioning (if barely) when the American political scientist Francis Fukuyama wrote an essay called "The End of History?" in which he argued that something more than the Cold War seemed to be coming to an end. What Gorbachev's twin reforms of glasnost (political openness) and perestroika (economic restructuring) signified, according to Fukuyama, was nothing less than the "unabashed victory of economic and political liberalism." He argued that the relentless extension of Western forms of consumer capitalism, combined with the utter exhaustion of any systematic alternative to liberalism, led to only one possible conclusion – the triumph of the West and the end of history as such. As Fukuyama put it, we had reached "the end point of mankind's ideological evolution and the universalization of Western liberal democracy as the final form of human government."

Despite the fact that the essay was larded with various caveats and rhetorical hedges (liberalism will only triumph "in the long run"; the question mark in the title), its publication was met with a gale of protest. By kicking the Soviet Union as it was down and trying to find firmer footing, "The End of History?" essay (and later, the best-selling book of the same title) was seen by many as embodying the worst elements of Western arrogance, a shallow manifesto for neoconservative imperialism. It certainly didn't help that the essay was published in the flagship neocon publication *The National Interest*. But there is far more to Fukuyama's argument than fin-de-communism gloating. The suggestion that we have reached the end of history offers a powerful lens through which to examine the events of the past twenty years, in particular those surrounding the fateful attacks of September 11, 2001.

The belief that history has a trajectory – a beginning, a middle, and an end – has been around in various forms since, well, the beginning of history. The particular version that Fukuyama is wrestling with comes out of the writings of Hegel, who thought that every civilization progresses through a series of stages of social organization: tribal, slave-owning, theocratic, aristocratic, and eventually, democratic egalitarianism. According to Hegel, the locomotive that pulls history through these successive stages is not the fight for survival and security, as Hobbes argues, nor is it the class struggle between the workers and the capitalists, as Marx believes, nor is it something as banal as the bourgeois desire for material comfort. No, the driving force of history is what he calls "the struggle for recognition."

What underlies the need for recognition is the sense that humans are essentially social creatures, and that our sense of dignity and self-worth is tied to how others see us. A man wants to be seen by his peers not as an inanimate object, or a mere animal, but as a person deserving of a certain amount of respect. In its earliest incarnations, the search for recognition results in a violent struggle for status as one man tries to establish dominance over another.

For Hegel, this battle for domination is the first authentically human act, because it marks the moment when men transcend their animal desires for self-preservation at any cost and show themselves willing to risk their lives in search of a greater dignity. But this is just the beginning, and the working out of history is the stepwise search for a way of satisfying both combatants' desire for mutual recognition on equal footing. The social order that accomplishes this goal is liberal democracy, and its achievement marks the end of history to the extent that it marks the end of the need for any further struggle for recognition.

To say that we have reached the end of history is not to say that there will be no more wars, no more jockeying for status between individuals or between states. It simply means that the ideological evolution of humankind has come to an end, insofar as there are no more imbalances or contradictions left to work out, politically, at the level of mutual recognition. Hegel actually thought that this point was reached in 1806, when Napoleon's army smashed the Prussian forces at the Battle of Jena. He believed this marked the triumph of the universal principles of the French revolution, and while there was obviously still a lot to do – slaves needed liberating, women needed enfranchising, priests needed disemboweling – these would be mere mopping-up operations, the inevitable working out of ideological principles that could not be improved upon.

When the cleanup is complete, human civilization will have reached "the universal and homogeneous state," a form of social organization that will be liberal and democratic in the political realm, and supported and fostered by a free-market driven consumer culture. In his 1989 essay, Fukuyama summarized the universal and homogeneous state as "liberal democracy in the political sphere combined with easy access to VCRs and stereos in the economic." Updated, we might say the end of history, the conclusion to the centuries-old struggle for recognition, is iPods and Xboxes for all.

There are plenty of cultural juxtapositions that herald the emergence of this new culture, such as the KFC that stands near the Sphinx in Egypt, the Starbucks that briefly took up residence in the Forbidden City in Beijing, or the McDonald's in the food court of the Louvre. But perhaps nothing quite says "End of History" with more brazen finality than a magazine ad a few years ago that featured a picture of Gorbachev sitting in the back of a limousine, looking out the window as his car drives past the last standing remnants of the Berlin Wall. Beside him on the seat is a Louis Vuitton bag. You could probably wring a thousand words out of just about any old photograph, but this one – shot by Annie Leibowitz and produced by the ad agency Ogilvy & Mather – is a Ph.D. thesis in waiting.

The suggestion that the endpoint of human development, the culmination of the ancient struggle for recognition, amounts to little more than the admixture of the Bill of Rights and Best Buy does not fill everyone's heart with joy. George Grant was not impressed with what he saw as the underlying nihilism of a liberal consumer society, and he argued that there can be no authentic (that is, no nontechnological, nonconsumerist, nonliberal) identity or culture in such a world, where difference exists only in the realm of private consumption. "Some like pizza, some like steaks; some like boys, some like girls; some like synagogue, some like the mass. But we all do it in churches, motels, restaurants, indistinguishable from the Atlantic to the Pacific."

Grant was writing in the late 1960s, but his concerns about the liberal state making authentic human life impossible remain extremely influential. Criticisms of this sort are the bread and butter of antiglobalization theorists, ecological activists, and antimodern declinists of all stripes who worry that the supposed cultural uniformity that has swept through North America and the West will eventually extend to the rest of the planet. In this view, liberal democracy is a genocidal force, eradicating local cultures and absorbing them into the nexus of technological liberalism and rampant consumerism.

Amid all of the hand-wringing, what has been almost completely lost is the fact that Francis Fukuyama himself was profoundly ambivalent about life at the end of history. The final, downbeat paragraph of his original essay is worth quoting in full:

The end of history will be a very sad time. The struggle for recognition, the willingness to risk one's life for a purely abstract goal, the worldwide ideological struggle that called forth daring, courage, imagination, and idealism, will be replaced by economic calculation, the endless solving of technical problems, environmental concerns, and the satisfaction of sophisticated consumer demands. In the post-historical period there will be neither art nor philosophy, just the perpetual caretaking of the museum of human history. I can feel in myself, and see in others around me, a powerful nostalgia for the time when history existed. Such nostalgia, in fact, will continue to fuel competition and conflict even in the post-historical world for some time to come. Even though I recognize its inevitability, I have the most ambivalent feelings for the civilization that has been created in Europe since 1945, with its north Atlantic and Asian offshoots. Perhaps this very prospect of centuries of boredom at the end of history will serve to get history started once again.

In this passage, Fukuyama confronts the stark consequences of the end of history, and concedes that it just may be everything its critics say it will be. He comes close to flinching, and he flags and even registers as justified a set of possible reactions that might keep us stuck in ideological comforts of history. What would it mean to "get history started once again"? In the context that he was writing in, it would mean one of three things. A return to either of the great totalitarian ideologies of the twentieth century, communism and fascism, or a resurgence of the sort of ethnic nationalism that the liberal cosmopolitans of the west fancy they have left behind.

The fascism that wracked Europe in the years leading up to the Second World War is gone, almost certainly forever, its energies successfully sublimated into the great project of building the European Union. There certainly was a surge of ethnic nationalism after the Berlin Wall tumbled, but more than anything, the nationalism in Eastern Europe in the 1990s was the result of long-term suppression of cultural and linguistic communities by the Soviets. The popular interpretation that it was a reaction against the end of history is almost certainly misguided, given that many of these countries virtually sued for admittance to the resolutely anti-nationalist European Union as soon as they were able.

Two decades after the demise of communism, it seems clear that what threatens to keep us in history is a combination of *nostalgia* and *boredom*. Nostalgia for a time when there were ideals worth fighting for, that required daring and courage and honor. Or for a time when one might be called on to die in the name of something more noble than protecting your stuff from a carjacking or a home invasion. And then a corresponding boredom with the spiritual residue of liberal democracy and consumer capitalism, a boredom so profound that it might be frightening. In which case, while we are right to be concerned about ongoing nostalgia in parts of Europe for the ideological comforts of the Cold War, the real worry might be the toxic blend of boredom and terror.

After the Soviet Union collapsed in 1991 and the fear and excitement of revolution had faded, the governments of the newly freed republics were faced with a single pressing question: what to do with all of those statues? A half-century of Soviet rule had left every major city from the Baltic to the Balkans littered with giant statues of Marx, Lenin, and Stalin, along with busts and sculptures of local communist apparatchiks. Some places tried to bury these eyesores out of sight and (hopefully) out of mind. For example, in 1991, the sixty-two–foot statue of Lenin that towered over Lenin Square in East Berlin was dragged off, cut into a hundred pieces, and buried in a secret place in the woods outside of town.

Hungary and Lithuania held competitions, soliciting bids from the public for ideas on what to do with the statues. In both cases, the curious answer was, *Let's build a theme park*. On a thirty-acre wooded area seventy-five miles from Vilnius, a sixty-year-old mushroom mogul and former wrestler named Viliumas Malinauskas has built the Soviet Sculpture Garden at Grūtas Park, better known to the locals as Stalin's World.

While traveling in Eastern Europe a few summers ago, my girl-friend and I paid the place a visit. At the entrance, there is a red cattle car of the sort they used to transport prisoners to the gulag. Once inside the park, we followed a winding boardwalk for a mile or so through the woods, along which some sixty-five Soviet-era statues stand peacefully amid the pines. At the end of the tour we sat at a picnic table next to an empty playground, eating kebabs while communist-era music blared from a tinny PA system. Slightly less weird (if only because it is less oddly located) is Statue Park, on the outskirts of Budapest. Opened in 1993, the park exhibits forty-two statues of various sizes, though not a single one of Stalin, since the only Stalin sculpture in Budapest was destroyed during the revolution of 1956. Despite the absence of such a major attrac-tion, Statue Park gets more than forty thousand visitors a year, making it one of the ten most-visited museums in Budapest.

These parks are elements of a growing commie tourism indus-try in Eastern Europe and parts of Asia. One thriving sector, let's call it "oppression tourism," gives visitors the experience of what it was like to live in the gulag or spend time in a Stasi cell. And so as a sort of sick side-dish to Grūtas Park, the Lithuanians recently opened something called Išgyvenimo Drama, Survival Drama in a Soviet Bunker. It provides experiences that include watching TV programs from 1984, trying on gas masks, learning the Soviet anthem under duress, eating typical Soviet food (with genuine Soviet tableware), and undergoing a concentration camp–style interrogation and medical check. To add to its KGB cred, all the actors involved in the project were originally in the Soviet army; some were even genuine interrogators.

A more playful scheme is the "Trabi tour," which allows tourists to explore communist-themed sights while driving around in a vintage Trabant. Noisy, slow, and a mobile environmental disaster, the two-cylinder, two-stroke Trabant is the Volkswagen of Eastern Europe, loved by the people not despite, but because of, its utter crappiness. In Berlin, a handful of companies will rent you a Trabant for a few hours and provide you with maps to all of the Soviet-era sights. One company's route takes you along the East Side Gallery – the longest piece of the Berlin wall that still stands – then down through Karl-Marx-Allee. As their brochure advertises, the Trabi-Safari will "take you back to the days of planned economy and real-life socialism."

While in Poland, we took a Trabi tour of Nova Huta, the Krakow suburb designed and built by Stalin as the very embodiment of communist ideology, "the ideal proletarian city." Our driver/guide was Mike, a slightly hyperactive law student from Krakow who met us outside our hotel wearing combat fatigues, a bleach-blond flattop, and a big grin. Mike told us that he started the company after it occurred to him that the old women selling vegetables at the local market were earning more than he would as a Krakow lawyer.

After giving us a brief lecture on the history of Nova Huta, Mike took us to a coffee shop that appeared to have been freezedried (along with the staff) in 1972. He showed us the broad Nova Huta streets that, according to rumor, were designed to accommodate Soviet tanks, drove us to the iron and steel factory that was built to employ the Nova Hutians, and walked us around the triangular housing projects, in one of which he had rented a showpiece apartment.

Mike had tricked the place out with Soviet-era appliances, electronics, and artwork. He played us communist music on a communist record player and served us communist coffee guaranteed to be the worst we had ever tasted. He did confess that, while the coffee wasn't quite as bad as the stuff that people had to drink back when the economy was centrally planned, the beans were indeed the absolute cheapest that he could find. So he was faking it, a bit. But in a way, that made the whole experience seem that much more authentic. After all, whatever else communism was, it was a magnificent bit of fakery visited upon millions by a relative handful of gangsters and fraud artists.

Communist kitsch is not exactly new. For years, visitors to Mao's mausoleum in Beijing have been able to walk away with key chains, pins, and cartons of Chairman Mao Memorial Filter Tips. You can buy Soviet military paraphernalia on the streets of Moscow, Prague, and even Budapest, despite the fact that the red-star symbol is officially banned in Hungary. Yet the growing commie tourism industry does force us to confront, once again, the problem of the easy ride that communism gets, in marked contrast with Nazism. As Anne Applebaum writes in the introduction to her book *Gulag*, it is odd that "while the symbol of one mass murder fills us with horror, the symbol of another mass murder makes us laugh."

It is an old problem, and you can take your pick of explanations. Some people point out that communism was an essentially inclusive and egalitarian ideology, in contrast with the exclusionary, racist hatred of the Nazis. Others suggest that the Soviets get off easy because so many writers and intellectuals in the West were (and are) leftists, while the Nazis supposedly get singled out as the purest example of absolute evil thanks to the efforts of a powerful Jewish lobby. At any rate, it isn't the sort of question that Mike was overly concerned about. When we asked him what other tour guide services he offered, he replied that while he takes tourists on the occasional trip to Auschwitz, he doesn't much like it. "This is more funky," he said, patting the dash of his Trabi.

While it takes a certain willful ignorance of the past to describe a totalitarian symbol as funky, as a business plan it makes a lot of sense. All travel is to some extent the quest for difference, and the more different (or exotic, or wild) a place is, the better. That is why we get annoyed when we go camping only to find the woods full of yahoos with radios and coolers of beer, and why it can be a bit depressing to fly halfway around the world only to see the people there wearing the same brands, eating at the same franchises, and listening to the same music as we get back home. Globalization has exacerbated the problem, and for the jaded world traveler, the urban centers of Poland, Hungary, and the Czech Republic just aren't that different anymore.

But what these places do have is their communist past, though some of the locals might prefer to pretend it never happened. When Stalin World opened, many Lithuanians objected, calling it an insult to the memory of those who were imprisoned, shot, deported, or otherwise repressed by the Soviets. Others replied that it was precisely the desire to ignore history and to remake society along utopian lines that made communism such a nightmare. They say that places such as Stalin World make it possible to remember communism by making it harmless, like drugged predators in zoos. The promotional material for Budapest's Statue Park confronts the issue head-on, emphasizing that the point of the institution is not to promote communist kitsch or propaganda, but to serve history. The park's motto is "no irony, memento."

Between the desire to forget and the imperative to remember, there is the realization that many tourists are coming to Eastern Europe precisely in the hope of feeling some faint shudder of what it must have been like under totalitarianism. A while ago, officials in Berlin were forced to disinter the buried statue of Lenin from its wooded grave and reinstate it, intact, in the city. The burial site had been discovered by souvenir hunters, who were digging up the pieces and chiselling off chunks of the red granite. Eastern Europe is still relatively poor and ill governed, and in places such as Poland and Hungary, the euphoria of EU accession has given way to a more pervasive gloominess. Tourism remains a major source of income, and a growing group of entrepreneurs can see that there is money to be made commodifying their former oppression. In the zero-sum world of global travel, a communist past is their competitive edge.

It would be a serious mistake, though, to assume that the need for tourist dollars is the only force driving the nostalgia for Europe's communist past. It is also propelled by a renewed desire for authenticity.

For a long time, it was assumed that any residual nostalgia for Stalin in Russia was a relic of the older generation, which would hold no appeal for increasingly liberalized and global-minded Russian youths. Yet instead of fading, Stalin's reputation within Russia has been steadily improving. In an article published in the January/February 2006 issue of *Foreign Affairs*, Sarah Mendelson and Theodore Gerber described the results of three surveys of Russian attitudes they conducted between 2003 and 2005. The results are pretty depressing. Fewer than half of all Russians said they would categorically reject voting for Stalin if he were in power today, while the country's youth appeared to hold "ambivalent, uncertain, or inconsistent views about the man."

This result has been confirmed in a number of polls since then, and Mendelson and Gerber attribute it to the failure of Russia to engage in a concerted de-Stalinization campaign. It probably doesn't help that in April 2005, then-president of Russia Vladimir Putin called the collapse of the Soviet Union "the greatest political catastrophe of the twentieth century" and then told a group of high school teachers that Russia's communist past was "nothing to be ashamed of." His government later commissioned some new textbooks aimed at providing a more "balanced" (that is, less critical) look at one of the great murderous tyrants of the twentieth century, comparing him to Otto von Bismarck, who fought to unify Germany in the nineteenth century. Similarly, Stalin's methods may have been harsh, but when it came to making the great people's omelet of the Soviet Union, what did it matter if a few million heads got scrambled along the way?

The simple fact is, there is a great deal of nostalgia not just in Russia but throughout the old Eastern bloc for the days of the Soviet empire, especially for Stalin's rule. Life may have been difficult, and there may have been a great deal of fear, but there is also a widespread sense that life was in many ways better – more purposeful, egalitarian, with a greater sense of community and social solidarity. The word for this is *Ostalgie*, a mashup of the German words for "East" and "nostalgia." Putin's neo-Stalinist ambitions may awaken memories of world-historical glories, but most aspects of *Ostalgie* are far more prosaic laments for the lost culture and daily life of the communist order.

The flagship document of the *Ostalgie* movement is the 2003 film *Good Bye Lenin!*, set in East Berlin in the aftermath of the fall of the Berlin Wall in 1989. It is the story of a small family that consists of a middle-aged woman named Christiane, who lives with her two kids, Alex and Ariane, their father having fled to the West decades before. Christiane has become an ardent supporter of the socialist government, and when she sees Alex roughed up by police during an antigovernment rally, she collapses into a coma. While unconscious, she misses the tremendous changes that occur, including free elections, the buildup to the unification of Germany, and the arrival of consumerism and capitalism.

When she emerges from her coma eight months later, Ariane has left school and taken a job at Burger King, while Alex now installs satellite television. But the doctors warn them that their mother's health is so fragile she must be protected from all shocks, so Alex and Ariane undertake a massive ruse: they keep the apartment dingy, wear their old, shabby clothes, and concoct fake news reports, all aimed at persuading their mother that the GDR is still intact. In the end, their desperate efforts to protect their mother's faith in socialism lead Alex and Ariane to reflect on their own attitudes toward unification and what capitalism has brought. Alex in particular finds himself thinking that things maybe happened too quickly, and that there was a lot about life in the old East Germany that was valuable. The sense of profound loss at the heart of the film continues to underwrite a substantial market in *Ostalgie*-related goods and experiences.

A number of entrepreneurs have resurrected some of the old brands of food and clothing, while state-produced TV programs have been reissued on DVD. The phenomenon is certainly controversial, and former GDR sex symbol and Olympic figure skating champion Katarina Witt annoyed a lot of people in 2003 when she signed on to host a television show celebrating the food, fashions, and occasional moments of levity in the old Soviet satellite.

Still, *Ostalgie* shows every sign of becoming an entrenched part of Eastern Europe's cultural landscape. While it is undoubtedly driven in part by a combination of youthful ignoramuses and bored irony hunters, there is also the darker sense among Eastern Europeans that they were sold a bill of goods, that the West has failed to deliver on the essential promise that Radio Free Europe used to broadcast across the Iron Curtain. In the late 1990s, a graffito started to appear throughout Eastern Europe: "We wanted democracy, but we ended up with the bond market." That bitter sentiment has only become more widespread over the past decade, as the momentum of post–Cold War euphoria and goodwill has slowed to a near halt. The gnawing sense is that liberal democracy and capitalism are not everything they were cracked up to be; at the very least, they were not obviously worth the exchange of community and statist order for individualism and rampant consumerism.

But deep down, everyone knows there is no going back. Only in Russia is the nostalgia for communism a potential threat, and even the worst-case scenario – that Putin succeeds in installing himself as de facto dictator-for-life – is not as big a problem as it may seem. Unlike the autarkic ambitions of Stalin and the other Soviet leaders, Putin's survival is tightly bound to Russia's ability to compete and prosper in the global economy. Globalization does not make future conflict impossible, but it does play a role in great-leader behavior modification.

For truly worrisome behavior, we need to look not at the places that have fought their way to the end of history only to find that it isn't everything it is cracked up to be, but at the ones who have taken a glance at what liberalism has to offer and are worried that it *is* everything it is cracked up to be. We need to look at radical Islam.

If acts are measured by their effects, then the destruction of the World Trade Center by Muslim jihadis on September 11, 2001, was undoubtedly an act of terror. But in addition to the visceral emotions of horror and vengeance that it spawned, the attack also created an instant cottage industry among opinion makers devoted to figuring out how to characterize what had happened and how to evaluate the nature of the threat. Al-Qaeda was no army, and it had the official backing of no state, so it was hard to say that the United States was now at war, at least not in the traditional sense. But it did not make sense to place al-Qaeda at the opposite end of the threat spectrum and treat them as common, albeit murderous, criminals. Neither the military nor the criminal option captured the true nature of the threat: nonstate actors who were nevertheless able to operate across national borders using highly lethal methods of "asymmetrical" warfare.

The Bush government settled for classifying suspected terrorists as "unlawful combatants," who were entitled to neither the constitutional protections of due process, nor the military protections of the Geneva Conventions, afforded to fighters who agree to play by the agreed-upon rules of war. The consequences of this decision now populate our working vocabulary with a minefield of new terms such as Guantanamo Bay, Abu Ghraib, waterboarding, special rendition, and warrantless wiretapping.

One of the more curious sidebars was the debate that arose over how to characterize the brand of Islamic terrorism that had been around for at least a decade but that erupted after 9/11, with subsequent attacks on Westerners in Madrid, Bali, London, and elsewhere. What should we call this violent jihadist ideology?

A number of commentators quickly picked up on the neologism *Islamofascism*. The implied parallels with the European fascist movements of the first half of the twentieth century served a double function: first, it made Islamic terrorist seem like an "existential threat" to the West on a par with the Third Reich, and, second, it underscored how irretrievably evil the perpetrators were. Just as there could be no reasoning or negotiating with Hitler, the fight against Islamofascism would be to the death. Unsurprisingly, the term proved particularly attractive to conservatives, although it is no coincidence that two of the most prominent advocates of its use, Christopher Hitchens and Paul Berman, are both former left-wing radicals transmogrified into liberal hawks.

The Islamofascist label has encountered strong resistance from the crew of counterculturalists, anti-imperialists, and campus liberals that makes up the antiwar left. For decades, *fascist* had served as this group's preferred insult for anyone with short hair, wearing a uniform, or holding a private sector job. As far as the left was concerned, the real ideological descendants of the Third Reich are to be found in the twin bastions of the American technocratic empire: Wall Street and the Pentagon – precisely the targets struck on 9/11. To call the attackers fascist gets the political valence of the actors exactly backwards.

Given the fairly strict partisan divide over the question, there is obviously a lot more at stake here than rhetorical point-scoring. As suggested by the early debate over how to classify the terrorists and their actions, words matter when they are embedded in relationships of power, authority, and violence. To call the brand of Islamic terrorism that leapt into our consciousness on 9/11 *Islamofascism* is to situate it in an ideological and military tradition with which we in the West are well familiar, and of which we are justifiably quite wary.

In *The Anatomy of Fascism*, historian Robert Paxton argues that while fascism was *the* major political innovation of the twentieth century, it resists definition. Unlike its companion totalitarian ideology Marxism, fascism was not based on a set of rational and potentially universalizeable principles. Instead, each version of European fascism – especially the purest forms, in Italy, Germany, and Spain – was tied to a particular national culture and institutional structure, and relied far more on charismatic leadership (the triumph of the will, you might say) than on theory. Mussolini in particular liked to brag that he was not tied to any specific doctrine, an intellectual promiscuity he attributed to self-salvation, writing, "With ideas as with women, the more you love them the more they make you suffer."

And so Paxton suggests that when looking to define fascism, we need to pay attention to each particular version, because it is the fact that each version is to some extent *sui generis* that makes fascism so different from other ideologies. To see the shape of the forest, we first need to look at the individual trees, comparing the manifestation of fascism in different places and at different times to see what pattern emerges.

In the end, what defines fascism in all guises is its uneasy, and frequently hostile, relationship to the three pillars of modernity – secularism, individualism, and consumerism. Communism and fascism are often lumped together as the twin "political religions" of the twentieth century, total systems designed to fill the spiritual vortex spinning in modernity's wake. But as Paxton concludes, fascism is a political movement marked by a rejection of liberal democracy, featuring an obsession with

- the primacy of the group or community over the individual
- a feeling of lost purity or unity
- a fear of decline due to alien cultural influences
- the worship of the charismatic leader over principles of reason
- a belief in the beauty and redemptive power of violence and death

This is pretty much the same definition proposed by Hitchens, though while Paxton flags the turn away from democracy as a key element of European fascism, Hitchens zeroes in on the cult of violence in pursuit of a lost purity as the defining characteristic. Hitchens argues that the effect of fascism was to transform politics from something based on reasoned argument over principles into a visceral aesthetic experience, and the most profound aesthetic experience for a fascist was murder.

In the trajectory of fascism, Paul Berman identifies a characteristic pattern that he calls "the modern impulse to rebel" and that emerged out of the French Revolution. This is not rebellion against authority but against freedom itself, rebellion in the name of submission to the kind of authority that had been undermined by modern liberalism. In each case, for every group, the pattern is the same. An initially pure and vibrant community is polluted from within and threatened from without, which leads to decline. The nostalgia for the past unity spurs an period of exterminating violence, which restores the community in all its glory, "without any of the flaws, competition, or turmoil that make for change and evolution."

Such is the ur-myth of fascism. In this view, the particular facts of the European variant, such as the turning away from democratic institutions, or the use of the state as the primary vehicle, are mere embellishments of the underlying dynamic, which is that fascism is an authenticity cult hostile to modernity and devoted to group purity and the aesthetics of violence.

———

As the world waited to see how the United States intended to avenge the destruction of the World Trade Center, many people were worried by the aggressiveness of George W. Bush's initial statements, especially the "with us or against us" rhetoric of the new "war on terror." But perhaps most galling was his declaration, the first Sunday after 9/11, that "this crusade, this war on terror, is going to take a while."

Bush's use of the term *crusade* to characterize the coming military response frightened a lot of people (mostly in Europe) who worried that he risked sparking a clash of civilizations. A typical reaction was that of Soheib Bencheikh, Grand Mufti of the mosque in Marseille, France, who said that "it recalled the barbarous and unjust military operations against the Muslim world," by Christian knights during the Middle Ages. These worries seemed borne out by subsequent attacks on Westerners throughout the world, from Bali (October 2002) to Madrid (March 2004) to London (July 2005).

These attacks did indeed herald a clash, but it is not between Islam and Christianity. Rather, it is a clash between Islam and *modernity*, and the relevant actors are not jihadis and crusaders, but jihadis and – for want of a better word – consumers. This is not about who is more worthy of worship, Muhammed or Jesus; it is about choosing between Muhammed and Lady Gaga.

The story of the rise of Islamic fundamentalism in the twentieth century has been written many times for Western audiences over the past decade, describing the emergence of al-Qaeda out of the Muslim Brotherhood and the roles played by men such as Hassan al-Banna, Sayyid Qutb, and Ayman al-Zawahiri. Qutb is the hinge figure: an intellectual and scholar who was driven to near panic by the way the modern values of secularism, individuality, and democracy had infected Islam. He railed against the "schizophrenia" of liberalism, which sought to keep religion subordinate through the separation of church and state. He believed that modernity drew a veil between man and God, and that technology and science had alienated humanity from the natural oneness with creation. Qutb feared what these Western ideas would do to Islam, and with some justification, given what Kemal Atatürk had done in secularizing the Turkish state. And so, as Lawrence Wright puts it in *The Looming Tower: Al Qaeda and the Road to 9/11*, Qutb's project "was to take apart the entire political and philosophical structure of modernity and return Islam to its unpolluted origins. For him, that was a state of divine oneness, the complete unity of god and humanity."

In the mind of Osama bin Laden, Qutb's rejection of Western rationalism became a hypertrophied revulsion for "America," which was jihadi shorthand for every aspect of the modern world, from politics (individualism, democracy, secularism) to business (globalization, trade, commerce) to pleasure (consumerism, alcohol, sex). As Wright points out, by turning their backs on virtually everything the twentieth century had to offer, the grand theorists of al-Qaeda left very little space for Islam to settle into a peaceful coexistence with modernity. The fundamentalists only had one place to go, and that was the past:

> Indeed, the man in the cave had entered a separate reality, one that was deeply connected to the mythic chords of Muslim identity and in fact gestured to anyone whose culture was threatened by modernity and impurity and the loss of tradition. By declaring war on the United States from a cave in Afghanistan, bin Laden assumed the role of an uncorrupted, indomitable primitive standing against the awesome power of the secular, scientific, technological Goliath; he was fighting modernity itself.

This description of al-Qaeda (and Islamic fundamentalism in general) as an authenticity movement devoted to the rejection of

American consumer capitalism helps explain one of the strangest formations on the post–9/11 intellectual landscape, which is the widespread sympathy from the left for the broad themes, if not the explicit theses, of al-Qaeda. That is, while hardly anyone was willing to come right out and celebrate the destruction of the World Trade Center and the killing of 3,000 civilians, a great many people found themselves nodding in agreement with bin Laden's general condemnation of Western culture. American capitalist imperialism was the problem, and while flying planes into towers might be excessive, well, blowback is unpredictable. You sow the wind and reap the whirlwind.

One of the most prominent American sympathizers was Ward Churchill, a humanities professor at the University of Colorado at Boulder who wrote an essay shortly after 9/11 in which he blamed the attacks on U.S. foreign policy and referred to the people working in the World Trade Center as "little Eichmanns." In Canada, barely three weeks after the attacks, a feminist professor named Sunera Thobani gave a speech at a conference in which she denounced the American-led global capitalist order, describing U.S. foreign policy as "soaked in blood."

It is no coincidence that both Churchill and Thobani – as well as the thousands of others who saw America as at least in some way the agent of its own destruction – fell back on familiar tropes linking capitalism, imperialism, and racism. A fun little quiz that has been floating around on the Internet for a while serves up a series of quotations on the subject of industrialization, consumerism, and the despoliation of nature. The reader is then invited to guess which document each passage is taken from, Al Gore's book *Earth in the Balance* or the *Unabomber Manifesto*. As anyone who has taken the quiz can attest, it can be surprisingly difficult.

What the quiz highlights is that there is a language that almost all of us instinctively fall back on when describing the downside of modern life. This "vocabulary of indictment" typically invokes themes of pristine nature "poisoned" by man's technological activities, the earth being scavenged or "raped" for its natural resources by our unnatural machines to fuel our "wasteful" and "voracious" appetite for consumer goods. That is, it relies heavily on the vocabulary of authenticity that is part of our common heritage as moderns.

Where we tend to differ is not in the vocabulary of indictment, but in our vocabulary of remediation – our suggestions for policies and institutions for coping with or fixing the various problems our activities cause. So what distinguishes a former vice-president of the United States from a homicidal mountainman is that the one thinks we should all drive Priuses and light our homes with compact fluorescent bulbs, and the other thinks that scientists should be blown up. It would be a pretty simple exercise to create a similar quiz with quotations from the writings or speeches of Sayyid Qutb or Osama bin Laden and running them beside passages from *The Long Emergency* by James Howard Kunstler or any given issue of *Adbusters* magazine.

The growing influence of the authenticity meme, especially (but far from exclusively) in the thinking of the left, has conditioned many people to see just about every major political or social problem as a consequence of modernity. Since the 1960s, leftists have been committed to the idea that some repressive and hegemonic system – variously referred to as capitalism, patriarchy, or the Man, but which we might as well just call modernity – is the single greatest threat to freedom. From this has emerged an intellectual template that divides the world into simple opposites like culture/nature, emotion/reason, hip/square, and so on. When the Cold War ended and the emerging global economy was met with a fierce backlash in places such as the Balkans and the Middle East, this template was broadened into what American political theorist Benjamin Barber calls "jihad vs. McWorld" – religious and nationalist identity-movements rebelling against cosmopolitanism, mass media, and consumerism.

Filtered through this template, it was no huge leap of logic to read the series of terrorist attacks against the West over the past decade as a natural and somewhat justified reaction against the progress of the universal and homogeneous state. Of course, terrorism and wanton mass murder are to be deplored, but this is merely a disagreement over tactics. Terrorism is just an extreme form of culture jamming, with suicide bombers the most committed members of the cause. The left trod a similar mental path when it came to the American plan to topple Saddam Hussein's Baathist regime in Iraq. "No blood for oil," went the chant; the assumption being that the only reason for invading was to make sure that the price of a gallon of gasoline remained less than that of a gallon of milk. As a consequence, the antiwar left found itself in a rather ridiculous position of excusing terrorism and mass murder, defending the sovereign rights of a tyrant, and selling out the rights of women to reactionary theologians.

The tragedy here is that the whole intellectual alliance between the antiwar left in the West and Islamic fundamentalists is based on a superficial equation (modernity equals bad) that masks far more complicated and divergent calculations. For many in the West, the whole problem with contemporary mass society is that it is inauthentic, but what makes it inauthentic is how packaged, homogenous, and conformist it all is. It lacks the naturalness, spontaneity, and uniqueness of the sort of culture that would promote true individual freedom and development.

The Islamicists agree that modern society is inauthentic, and they even agree that consumerism is a big part of the problem. But for the Islamicists, what makes modernity inauthentic is not that it is homogeneous and conformist, just the opposite. For bin Laden and his followers, the problem with the West is how creative, spontaneous, and individualistic it is. So much so, that it impedes the creation and sustenance of an authentic Muslim community, which requires a great deal of conformity of thought, of worship, of dress, and of habit.

Nowhere does this elementary confusion manifest itself more painfully than in the writings of Naomi Wolf, best known for *The Beauty Myth*, which became the early 1990s bible of campus feminism. In it she claimed that the idea that some women are more beautiful than others is nothing but a "collective reactionary hallucination" concocted by "the social order." The book is mostly just reheated counterculturalism, full of statements to the effect that beauty serves "nothing more exalted than the need of today's power structure, economy, and culture to mount a counteroffensive against women." As she argues, the social order creates a series of beauty archetypes – such as supermodels, Barbie, and *Playboy* bunnies – that represent the ideal of female beauty. These "formulaic" images are "endlessly reproduced" in the media, primarily through advertising. Women, in turn, are brainwashed by these images, and strive to conform to the archetypes, which reproduces the power structure.

Wolf ends up positing the idea that there are two types of individuality – a "conformist" individuality that is endorsed by the prevailing power structure, and an "authentic" individuality that rejects and even subverts the competitive needs of that structure. Ultimately, what women need to do, she says, is free themselves entirely from the false competition of the beauty myth and create a new "pro-woman," nonhierarchical definition of beauty that would celebrate each different woman in her own uniqueness and individuality.

Seventeen years later, Naomi Wolf seems to have finally realized that there is more to getting out of the beauty game than simply shrugging it off, and after spending some time traveling in Morocco, Jordan, and Egypt in 2008, she had a revelation about the

Koranic requirement that Muslim women cover their heads. There are varying degrees of modesty, ranging from the simple headscarf, or hijab; through the cloaklike chador; all the way to the burka, a head-to-toe outfit that leaves only a slit for the eyes.

One interpretation of this enforced veiling of women is that it is a form of sexual repression that is only the most obvious manifestation of Islam's astounding sexism. Yet on her trip, Wolf discovered that Islam's requirement that women cover up is actually a sign of the culture's healthy sense of what is public and what is private, "of what is due God, and what is due one's husband." She came to appreciate the way the various coverings serve to actually liberate women from the "intrusive, commodifying, basely sexualizing Western gaze." She began to worry that our hostility to the veil, the chador, and the burka has blinded us to our own markers of the oppression and control of women (by which she means beauty industry staples such as makeup and high heels).

And so she tried an experiment. She pulled on a shalwar kameez (a traditional dress) and a headscarf and went for a stroll in a Moroccan bazaar, and was surprised at what a liberating experience it was:

> Yes, some of the warmth I encountered was probably from the novelty of seeing a Westerner so clothed; but, as I moved about the market – the curve of my breasts covered, the shape of my legs obscured, my long hair not flying about me – I felt a novel sense of calm and serenity. I felt, yes, in certain ways, free.

Naomi Wolf took a lot of heat for this article. Some of her critics were liberal feminists in the West, but the most vicious reactions came from liberal Muslim women who found themselves gasping for air at what they saw as her breathtaking ignorance. In particular, she was accused of ignoring the question of choice, since the hijab or the burka isn't just something a woman pulls on one morning when she doesn't feel like getting dolled up to go shopping, the way Hollywood starlets wear yoga gear to grab their coffee. The covering of women is just one aspect of the thoroughly patriarchal cultural attitudes that underwrite atrocities like the acid routinely thrown in the faces of Afghan girls who dare go to school or the thirteen-year-old Somali girl who in October 2008 was raped by three men and then stoned to death for adultery by her family.

These are legitimate objections, and Wolf completely deserved the dusting she received. But what got lost amid the outrage was the remarkable way in which her argument completes her journey away from the positions of *The Beauty Myth*. She has accepted that there is no way of casually shrugging off the competitive beauty dynamic that the social order supposedly requires and simply embracing a more authentic and nonhierarchical individualized form, since it doesn't exist. As women have been telling men for centuries, women don't dress up for themselves, and they certainly don't dress up for their men. They dress up for other women.

The only way out of the beauty game then is to repudiate competitive individuality entirely and seek refuge in the sort of conformity and collectivism you find in Islam. For a Western woman traveling in a hot country, a day spent without having to worry about her hair may certainly seem liberating, but Wolf managed to yank the hijab out of its deep religious context and drain it of all its barbarous connotations. This isn't political commentary, it is a form of authenticity tourism, and in her jaunt through the Moroccan bazaar, Wolf learned as much about what it means to be a Muslim woman as the visitors to Išgyvenimo Drama theme park in Lithuania learn about what it was like to be a prisoner of the KGB.

It probably isn't fair to single out Naomi Wolf for criticism. The attacks of 9/11 wrong-footed just about everyone in the ideas business, introducing strong and conflicting crosscurrents into the mainstream ideological divides into which most of us were floating comfortably. The destruction of the World Trade Center had an especially disorienting effect on novelists, those unacknowledged legislators who suddenly found their imaginative talents less in demand. Faced with the terror of what had happened and total uncertainty about what might be next, the government and the public alike turned to historians, political scientists, and experts in the geopolitics of the Middle East and central Asia.

Some of the braver (or more foolish) writers tried to somehow out-write the terrorists, to get on top of the fray not through insight but through desperate and breathless turns of phrase. The results were almost uniformly embarrassing, the low-water mark reached by Don DeLillo in an essay entitled "In the Ruins of the Future," published in the December 2001 edition of *Harper's* magazine. It is a confused and contradictory mess of stream of consciousness, the tone is driven by ponderous non sequiturs and short fragments of thought.

Martin Amis was another of the brave ones, though; he was sufficiently self-aware to realize that the basic problem is that writing fiction is ultimately not a serious business, and when the times turn serious, people are not likely to turn to novelists for guidance. That is why an inordinate number of fiction writers took up journalism after 9/11. They were, as Amis puts it, "playing for time" while they struggled to figure out the proper place for imaginative writing in a world where a handful of murderous Islamist thugs had out-imagined just about everyone.

Amis struggled more than most to find his footing, and in his collection of post–9/11 writings he jerks from style to style in an inconsistent mix of short commentary, book reviews, a couple of longer essays, and a handful of atrocious attempts at terror fiction. The most interesting piece in the lot is the essay "Terror and Boredom: The Dependent Mind," which riffs off the tedium of standing in the security line in an airport and tries to situate the terrorist's lust for death on the expansive stage of end-of-history ennui.

Amis spends a lot of time steering around the point here, but he eventually arrives at the claim that a world where the terrorists had won would be "a world of perfect terror and perfect boredom, and of nothing else – a world with no games, no arts, and no women, a world where the sole entertainment is the public execution":

> The age of terror, I suspect, will also be remembered as the age of boredom. Not the kind of boredom that affects the blasé and the effete, but a superboredom, rounding out and complementing the superterror of mass suicide-mass murder. And although we will eventually prevail in the war against terror . . . we haven't got a chance in the war against boredom. Because boredom is something that the enemy doesn't feel.

There is actually a depth of insight to be plumbed here, but Amis never takes the terror/boredom contrast beyond the smug satisfaction of having arrived at a clever juxtaposition. Instead, he veers off into a criticism of the handling of the war in Iraq, and concludes the essay with an endorsement of Liz Cheney's (daughter of Dick Cheney, and a lawyer) silly idea that the billions of dollars spent invading Iraq would have been better spent funding "consciousness raising" among Muslim women. Which is too bad, because that simple opposition of terror and boredom provides any number of avenues of approach. Whose terror? What sort of boredom? Amis simply assumes that the terror at issue is ours and that the boredom would be the profound tedium of an independent mind tethered against its will to a totalitarian religious system.

But there is another way of looking at it that is more sympathetic to the ambitions of the fundamentalists, and which considers their terror, their boredom. This isn't the boredom of a long bus ride or a few hours in a doctor's waiting room; it isn't even the world-weariness one feels from having spent too long in one place, or too long on the move. Instead, it is the boredom of the abyss, the boredom that Fukuyama conceded would be one consequence of the end of history, the terrifying feeling that in a world where everything is possible, there may be nothing of value. What the fundamentalist fears is the terror of the mind cast loose from all possible moorings, the pure liberal intellect that floats in a limitless sea of possibility but that can find nowhere worth anchoring.

This is everyone's problem. Eventually, each of us has to look in the mirror and ask, "What is worth doing?" or "What is meaningful?" or "What is sacred?" These are all versions of the same question, and what they amount to is, "Who am I?" Islam, like all religions or ideologies, gives a ready-made answer to that question. But modernity sweeps away all previous answers, undermines any notion of the sacred. And so the liberal answer to these questions is, "Nothing" – or, slightly better, "It's up to you." This can be terrifying, and while we can – indeed must – condemn those who turn their backs on modernity and seek refuge in nostalgia and violence, we must also recognize that our own solution, the confused and self-defeating search for something called authenticity, is itself nothing more than a hoax.

—

PROGRESS, THE VERY IDEA

THE QUEST FOR AUTHENTICITY IS ABOUT SEARCHING FOR meaning when all the traditional sources no longer have any sound, rational justification. This book is an exploration of the quintessentially modern attempt at replacing these sources with something more acceptable in a world that is not just disenchanted but also socially flattened, cosmopolitan, individualistic, and commercialized. And so the search for authenticity is motivated by a visceral reaction to secularism, liberalism, and capitalism, and the sense that a meaningful life is not possible in the modern world, that all it offers is a toxic mix of social-climbing and alienation. Absent from our lives is any sense of the world as a place of intrinsic value, within which each of us can lead a purposeful existence. And so we seek the authentic in a multitude of ways, looking for a connection to something deeper in the jeans we buy, the food we eat, the vacations we take, the music we listen to, and the politicians we elect. In each case, we are trying to find at least one sliver of the world, one fragment of experience, that is innocent, spontaneous, genuine, and creative, and not tainted by commercialization, calculation, and self-interest.

It is obvious to anyone paying attention that the civilization we call "the West" (which includes the industrialized and postindustrial economies of North America, Europe, and parts of Asia) has its share of problems. There remains plenty of social and economic inequality, lots of crime and corruption, not to mention illness, poverty, pollution, and other forms of environmental degradation. As a symptom of how bad our social malaise seems to be getting, a study revealed that by 2005 one American in ten was on antidepressants, a figure that doubled over the preceding decade.

Yet it is hard to square the bleak picture of modern life with the undeniable fact that by any reasonable measure of human development things are better here than anywhere else, and they are better here today than they have ever been. We live longer, healthier lives, our air and water are cleaner, and we have almost universal access to plumbing, heating, electricity, medicine, television, Internet, and other utilities and services. There is more entertainment, music, film, radio, television, news, and other information easily available, and the market provides consumer goods and services catering to every imaginable taste and lifestyle. In short, life for the citizens of developed liberal democracies is, on the whole, more enjoyable than it has ever been on earth. So what seems to be the problem?

In his book *Unpopular Essays*, philosopher Bertrand Russell notes that the misfortunes that can befall humanity can be sorted into two broad categories: things that are inflicted by nature and things that are inflicted by humans. For most of history, a great deal of suffering was due to natural causes, such as famine and disease, but as we have developed in knowledge and skill, the class of harms inflicted by other humans has come to occupy a greater percentage of the total. Put crudely, there is less famine but more war, and as a result we end up with the impression that "nature" is relatively benign, while "civilization" is increasingly a threat.

And so modernity is to some extent a victim of its own success. Only people who live long and relatively comfortable lives have the luxury of wondering what it's all about. Only moderns are in a position to discover that shopping and voting do not truly satisfy the deepest needs of the human spirit. Our lives might be pleasant but they aren't meaningful, and the ongoing search for the authentic is ultimately driven by the sense, felt by millions of people, that there has to be more to life than this.

This is a problem that philosopher Friedrich Nietzsche flagged in his root-and-branch critique of the morality that he saw as underpinning liberal democracy. In the last chapter, we looked at the idea that the triumph of liberalism represents the moment when one of history's most basic drives, the search for recognition, reaches its final expression: each and every person receives full and equal recognition in their uniqueness and individuality. For Nietzsche, the figure that emerges at the end of history, the supposed hero of the liberal state, is what he calls "the last man."

But Nietzsche saw nothing heroic in these last men who value democracy, equality, freedom, and security above all else. As far as he was concerned, the emergence of the last man represents the final triumph of the morality of the slaves, of men who would rather save their own skins than lose out in a fight for domination. Liberalism, according to Nietzsche, is the victory of the herd, of the weak over the strong, of a group whose morality triumphs not because it is better but because there are just far more of them around. There are no real leaders in a liberal state, only the "citizen" who has given up any "prideful belief in his or her own superior worth in favor of comfort and self-preservation."

Nietzsche's contempt for the morality of liberal democracy is what led him to argue that only in aristocratic societies could there be true excellence and achievement, because only in such societies are there people willing to take risks that make them worthy of recognition. These people are not content to be recognized as equals; they instead want to establish themselves as superior, to stand out from the crowd. This desire for superiority is more than just the stereotypical desire for conquest and exploitation, it is the underlying motivation for everything that is worth doing or having in life, from great symphonies to "painting, novels, ethical codes, or political systems."

This helps explain why the search for the authentic is a trap from which it is almost impossible to extricate ourselves, given the terms of the search itself. It begins with Nietzsche's simple but penetrating insight that when recognition is universal, it becomes trivialized to the point of worthlessness. To put it bluntly, if everyone is recognized, then no one is. Recognition is inherently a form of what sociologist Pierre Bourdieu calls "distinction" – something that is shot through with power relations and judgments about higher and lower, better and worse. If we transpose the terms of debate from *recognition* to *authenticity*, the claim is that if everything is authentic, then nothing is because *authenticity* is a contrastive term that gets whatever force it has through answering the question "Authentic as opposed to what?"

In the end, authenticity is a positional good, which is valuable precisely because not everyone can have it. The upshot is that, like the earlier privilege given to the upper classes, or the later distinction gained from being cool, the search for the authentic is a form of status competition. Indeed, in recent years authenticity has established itself as the most rarified form of status competition in our society, attracting only the most discerning, well-heeled, and frankly competitive players to the game.

Any status hierarchy is socially pernicious when it is used to allocate scarce goods and resources on the basis of arbitrary or unearned qualities. It is good to be the king, and almost as good to be a prince, or a duke, or a count, and on down the aristocratic chain. But not all forms of status are illegitimate: higher education is a status hierarchy that helps allocate wealth and privileges, yet for many people, the fact that the education system is for the most part a meritocracy makes it a fair, just, and even democratic form of status competition.

What makes the quest for authenticity a socially destructive form of status-seeking is the way it takes the misguided critique of mass society that has motivated the quest for "cool" for the past forty years and blows it up into a sweeping and even more wrong-headed critique of the entire modern world. The critique of mass society led many people, especially on the left, to be wary of the building blocks of social organization, including basic social norms, the bureaucratic state, and the legal system. The critique of modernity goes a giant step further with its indictment of the entire scientific, legal, and political foundations of liberal democracy and the culture in which it flourishes.

This manifests itself in degrees, from those who go through the motions of being pious about authenticity to those who are positively fundamentalist about it. Some engage in a relatively inert nostalgia for a nonexistent past, others fetishize the exotic, still others find hope in collective struggles like nationalism or even jihad. But the collective, paradoxical result is that the search for the authentic almost always ends up contributing to the very problems that we are trying to escape. This is because we are caught in the grip of a false ideology about what it means to have authentic experiences, to be an authentic self, to lead an authentic life.

David Suzuki is a Japanese Canadian who started his career as a geneticist, became a wildly popular science journalist as host of the television show *The Nature of Things*, and then later in life dedicated his energies to the environment. He has been awarded twenty-two honorary degrees and is an internationally recognized expert on climate change.

But at some point along the line, Suzuki lost touch with reality and common sense. In a column he wrote for a Canadian newspaper on the eve of his seventy-third birthday, he wondered whether, despite the scientific and technological advances of the twentieth century, the world is actually a better place today than when he was born. In order not to leave the reader in suspense, he cut to the conclusion: "Reflecting on what we leave to our grandchildren, I have to answer with a resounding no!"

Suzuki was willing to concede that there were a few good developments over the decades, like penicillin and his computer, but that's about as far as he would go. Otherwise, the modern world has been a disaster:

> Yes, our world now provides a cornucopia of wondrous consumer goods. But at what cost? When I was a child, back doors would open at 5:30 or 6 o'clock as parents called kids for supper. We were out playing in grassy fields, ditches, or creeks. We drank from rivers and lakes and caught and ate fish, all without worrying about what chemicals might be in them. . . . The population has tripled since then. Each of us now carries dozens of toxic chemicals embedded within us, cancer has become the biggest killer, and we have poisoned our air, water, and soil.

It is worth noting that David Suzuki was born in 1936. What lay ahead was not just three more years of economic depression and then a ridiculously destructive world war. There was also the internment camp. He doesn't mention it in his article, but the Rousseauian idyll of his childhood was interrupted by a decision by the Canadian government, in 1942, to confiscate his family's dry cleaning business, put his father in a labor camp, and intern the rest of the family in a camp in the interior of British Columbia. In 1988, the government apologized for this treatment of Japanese Canadians and offered cash payments to anyone who was affected by it. This apology, and the shame that motivated it, is evidence of clear moral and social progress, though it too goes unacknowledged.

We can skip quickly over the rest of the rebuttal to Suzuki's argument, such as the reason the population has tripled is because instead of one in five children dying at birth in the United States (as was the case in 1900), today that figure is less than one in a hundred. We can also cite the figures showing that poverty and malnutrition were far more widespread in Canada in 1936 than in 2009, and argue that while environmental problems persist, there have been many great successes as well.

There's no point dwelling on any of this, though, because this is the authenticity hoax in full throat: a dopey nostalgia for a non-existent past, a one-sided suspicion of the modern world, and stagnant and reactionary politics masquerading as something personally meaningful and socially progressive. At the heart of the authenticity hoax is the assumption that what is good for me must also be good for society, good for the planet, and just plain old Good. It assumes that what is spiritually satisfying will also be morally praiseworthy, and that if you find your way to the first, the second will follow as a matter of course.

But as I've argued throughout this book, we have no right to make this assumption. It may on occasion happen that things we find personally meaningful are also socially beneficial, but there is no necessary connection between the two. As we've seen, more often than not, the search for the authentic quickly turns into a socially regressive arms race that only intensifies the very competition from which we're supposedly trying to escape.

In order to see ourselves clear of the authenticity hoax, we need to come to terms with the modern world and accept that the last 250 years or so has not been a tragic mistake. At the very least we have to concede that while there has been a trade-off, losses to balance against the benefits, on the whole it would be a mistake to want to pull the plug, put the wagon train in reverse, and head back into the nostalgic comforts of history.

But we can go further than that. Coming to terms with modernity involves embracing liberal democracy and the market economy as positive goods. That means not just conceding that they are necessary evils, but that they are institutions of political and economic organization that have their own value structure, their own moral foundations, which represents a positive step away from what they have replaced. So even if it were possible, it would be wrong to turn our backs on the market. Nor should we pine for a perfectionist social order that would undermine the rights and freedoms that allow for the flourishing of countless lives that in other places, or at other times, would have been subject to resentment and repression.

The search for the authentic has fooled millions of well-intentioned people over the years, leading them into both sin and betrayal. It is a sin because it displays an utter lack of faith in humanity, believing that we will inevitably abuse the gifts of freedom, knowledge, and power, and become the agents of our own destruction. It is a betrayal of modernity and of the liberal ideals that have breathed life and hope into human progress for the past quarter millennium.

Progress is a stuffy old word, employed primarily by squares and ironists. But perhaps it is time to rehabilitate the very idea of progress: not the blind conviction that things are getting better all the time, but the simple faith that even when humans encounter obstacles, we'll figure things out, through the exercise of reason, ingenuity, and goodwill. Faith in progress is nothing more, and nothing less, than faith in humankind, and if there is one thing we ought to be nostalgic for, it is for a time when progress was something that self-described "progressives" actually believed in. For too long now they've been wallowing in an inert philosophy that has done considerable damage to the search for social justice and spiritual comfort.

Ludwig Wittgenstein said that the trick to doing philosophy is knowing when to stop asking the questions that lead us awry. When it comes to the modern search for authenticity, the irony is that the only way to find what we're really after might be to stop looking.

ACKNOWLEDGMENTS

This book took much longer to write than it should have. I owe a huge debt of thanks to my agent, Michelle Tessler, who liked the idea from the very start and who continues to stand by me. I'd also like to thank Doug Pepper of McClelland & Stewart, who along with Chris Bucci saw fit to bring me into the M&S fold.

I've had the privilege of working with two excellent editors, Trena White of McClelland & Stewart in Canada, and Ben Loehnen of HarperCollins in the United States. Between them they provided the perfect combination of encouragement and criticism, and they pushed me to make the arguments sharper while dragging me back from some unwise directions.

Heather Sangster did an outstanding job copy-editing the manuscript, and she made plenty of helpful suggestions for improving both the content and the writing. This book's unofficial editor is Sarmishta Subramanian of *Maclean's* magazine. I run my flakiest ideas by her first, and she always makes me sound a lot smarter than I am. Laura Drake was the first person to read the manuscript from beginning to end, and her helpful feedback came at a crucial time. The better jokes in the text are probably Laura's.

My interest in the question of authenticity goes back to my time as a graduate student in philosophy at the University of Toronto. Those remain some of the most stimulating years of my life, largely thanks to the privilege I had of working with professors Ronald de Sousa, Mark Kingwell, and Joseph Heath. I've helped myself to their ideas throughout this book, as they'll discover once they read it. Joe in particular has become a friend and mentor, but he remains my most important teacher as well.

Finally, I would like to thank Elizabeth Wasserman, who was present at the creation and at every stage along the way. Liz always knows what I am trying to say better than I do myself, and her influence is on every page. For her love, faith, and encouragement I'll always be grateful.

NOTES

INTRODUCTION

2   *"The pirates must not be allowed to destroy our dream"*: Colin Freeman and Mike Pflanz, "Rescued French yacht captain Florent Lemaçon may have died in friendly fire." *Daily Telegraph*, April 11, 2009.

5   *"greed, overspending, obsession with luxury and brands"*: John Zogby, *The Way We'll Be: The Zogby Report on the Transformation of the American Dream* (New York: Random House, 2008), 150. All references to Zogby in the following paragraphs are from Chapter 6: "One True Thing: Searching for Authenticity in a Make-Believe World" of this book.

7   *"It was a moment of almost spiritual doubt"*: David Boyle, *Authenticity: Brands, Fakes, Spin and the Lust for Real Life* (London: Harper Perennial, 2003), 2.

7   *"dominated by spin doctors, advertising, virtual goods and services"*: Ibid., 3.

10   *"has had a moment of self-transcendence"*: Lionel Trilling, *Sincerity and Authenticity* (Cambridge: Harvard University Press, 1971), 4.

11   *But right from its opening sentence*: Harry G. Frankfurt, *On Bullshit* (Princeton: Princeton University Press, 2005), 1.

11   *"has become part of the moral slang"*: Trilling, *Sincerity and Authenticity*, 93.

CHAPTER I

18   *Philosophy, he told them*: Plato, "The Apology," in *The Trial and Death of Socrates*, trans. G.M.A. Grube (London: Hackett, 2001), 38a.

20   *Modernity is what Marshall Berman*: Marshall Berman, *All That Is Solid Melts Into Air: The Experience of Modernity* (New York: Penguin, 1982), 15.

22   *This was the worldview in which Socrates*: See the discussion in chapter 2 of Charles Guignon, *On Being Authentic* (New York: Routledge, 2004).

25   *Pope John Paul II employed both of these methods*: His Holiness Pope John Paul II, *Address to the Pontifical Academy of Sciences*, October 22, 1996, www.christusrex.org/www1/pope/vise10-23-96.html.

27   *"That the nurture of all creatures is moist"*: Aristotle, "Metaphysics," in *A New Aristotle Reader*, ed. J.L. Ackrill (Princeton: Princeton University Press, 1987), 983 b23–27.

28 *For Weber, the commitment to science:* Max Weber, *From Max Weber: Essays in Sociology,* trans. and ed. H.H. Gerth and C. Wright Mills (New York: Oxford University Press, 1946), 139.

30 *The deep connection between the rise of the modern state:* Larry Siedentop, *Democracy in Europe* (New York: Columbia University Press, 2001), 89.

32 *"The very idea of the state":* Ibid., 83.

37 *"Rights are best understood as trumps":* Ronald Dworkin, "Rights as Trumps" in *Theories of Rights,* ed. Jeremy Waldron, (Oxford: Oxford University Press, 1984), 153.

38 *Smith argued that each individual:* Adam Smith, *The Wealth of Nations* (New York: Barnes & Noble Books, 2004), 14.

39 *the first great consumer revolution:* For an excellent survey, see Judith Flanders, *Consuming Passions: Leisure and Pleasure in Victorian Britain* (London: Harper Perennial, 2007).

39 *everyone who was anyone:* Colin Campbell, *The Romantic Ethic and the Spirit of Modern Consumerism* (London: Blackwell, 1987), 22.

40 *"Innovation was catching":* David Landes, *The Wealth and Poverty of Nations: Why Some Are So Rich and Others So Poor* (New York: W.W. Norton & Co, 1999), 191–192.

42 *"The bourgeoisie, during its rule of scarce one hundred years":* Karl Marx, *The Communist Manifesto,* trans. F.L. Bender (New York: W.W. Norton, 1988), 59.

43 *"anyone who does not actively change on his own":* Berman, *All That Is Solid Melts Into Air,* 94–95

44 *"Constant revolutionizing of production":* Marx quoted in Berman, *All That Is Solid Melts Into Air,* 94.

CHAPTER 2

51 *"I was questioned and denied":* Jean Starobinski, *Jean-Jacques Rousseau: Transparency and Obstruction* (Chicago: University of Chicago Press, 1988), 8.

53 *"consider myself as having no hands":* René Descartes, "Meditations on First Philosophy," in *The Philosophical Writings of Descartes Vol. II,* trans. Cottingham, Stoothoff, and Murdoch (Cambridge: Cambridge University Press, 1984), 15.

53 *"I am, I exist is necessarily true":* Ibid, 17.

57 *"I see an animal less strong than some":* Jean-Jacques Rousseau, "Discourse on the Origins and Foundations of Inequality Among Mankind," in *The Social Contract and The First and Second Discourses,* ed. Susan Dunn (New Haven: Yale University Press, 2002), 90.

57 *"One of them interests us deeply":* Ibid., 84.

58   *"Do good to yourself"*: Ibid., 108.

58   *"after enclosing a piece of ground"*: Ibid., 113.

59   *In becoming aware of how they compare:* Ibid., 115.

60   *"Men no sooner began to set a value upon each other"*: Ibid., 118–119.

62   *"I have received, sir, your new book against the human species,"* Voltaire, *Voltaire's Correspondence*, ed. Theodore Besterman (Geneva: Institut et Musée Voltaire, 1957), 230.

62   *"the tribal world was morally transformed"*: Roger Sandall, *The Culture Cult: Designer Tribalism and Other Essays* (Oxford: Westview Press, 2001), 39.

63   *Deschamps proposed a world where intellectuals would be banned:* Ibid., 48.

63   *"despotic chiefs, absurd beliefs, revolting cruelty, appalling poverty"*: Ibid., 39.

64   *"This last place on Earth"*: Nicholas Wade, *Before the Dawn: Recovering the Lost History of Our Ancestors* (New York: Penguin, 2006), 86.

65   *"creates a pervasive atmosphere of ambiguous make-believe"*: Sandall, *The Culture Cult*, 47.

67   *For a* New Yorker *article, writer Ben McGrath:* Ben McGrath, "The Dystopians," *The New Yorker*, January 26, 2009.

68   *"Our perception of what we are"*: The Prince of Wales, "The Modern Curse that Divides Us from Nature," *The Times of London*, November 27, 2008, www.timesonline.co.uk/tol/comment/columnists/guest_contributors/article5240226.ece.

71   *And so we use our art to destroy New York:* Max Page, *The City's End: Two Centuries of Fantasies, Fears, and Premonitions of New York's Destruction* (New Haven: Yale University Press, 2008), 9.

72   *The central concern of Rousseau's philosophical project:* The remainder of this chapter is heavily indebted to the excellent discussion of Rousseau's "Romantic turn" in Charles Guignon, *On Being Authentic* (New York: Routledge, 2004), 49–78.

73   *"I truly am what I feel myself to be"*: Ibid., 68–69.

CHAPTER 3

77   *"I want to believe there are things going on"*: Siri Agrell, "Does the Artist's Story Affect the Art?" *National Post*, September 12, 2006, www.canada.com/nationalpost/news/toronto/story.html?id=592836e9-6bcc-4a96-aab5-2910b1a0da7c.

83   *as far back as the Phoenicians:* Thomas Hoving, *False Impressions: The Hunt for Big-Time Art Fakes* (New York: Touchstone, 1996), 24.

87   *"He sometimes distressed his colleagues"*: Ibid., 20.

88   *"is the manner or personal style of the artist"*: Francis V. O'Connor, "Authenticating the Attribution of Art," in *The Expert Versus the Object:*

Judging Fakes and False Attributions in the Visual Arts, ed. Ronald D. Spencer (New York: Oxford University Press, 2004), 10.

89 *"Just as most men, both speakers and writers"*: Ibid., 33.

92 *"The ship wherein Theseus and the youth of Athens"*: Plutarch, *Theseus*, trans. John Dryden, Internet Classics Archive, www.classics.mit.edu/Plutarch/theseus.html.

93 *"Artists and conservators have different opinions"*: Carol Vogel, "Swimming with Famous Dead Sharks," *The New York Times*, October 1, 2006, www.nytimes.com/2006/10/01/arts/design/01voge.html.

94 *The case eventually went to court*: Ronald D. Spencer, "Authentication in Court," in Ronald D. Spencer, *The Expert Versus the Object*, 198.

96 *Benjamin argues that there is a straightforward answer*: Walter Benjamin, *Illuminations: Essays and Reflections* (New York: Schocken, 1969), 20.

98 *"The idea that there is some special magic"*: Robert Hughes, "Day of the Dead," *The Guardian*, September 13, 2008, www.guardian.co.uk/artanddesign/2008/sep/13/damienhirst.art.

99 *"there is almost nothing you can buy for £1 million"*: Don Thompson, *The $12 Million Stuffed Shark: The Curious Economics of Contemporary Art* (Toronto: Doubleday, 2008), 16.

100 *There has been a lot of talk recently about the rise*: Chris Anderson, *Free: The Future of a Radical Price* (New Yorker: Hyperion, 2009).

101 *A more delightful example of the attention economy*: S. Mitra Kalita, "Not-So-Easy Listening: It Takes a Trek to Hear This Track," *The Wall Street Journal*, June 12, 2009, A1.

CHAPTER 4

103 *"shiny, fabricated world of spun messages"*: Bill Breen, "Who Do You Love?" *Fast Company*, May 2007, 82.

104 *Being able to play the authenticity game*: James Gilmore and Joseph Pine, *Authenticity: What Consumers Really Want* (Cambridge: Harvard Business School Press, 2007).

104 *The Levi's story is pretty well known*: This section was inspired by a correspondence with Sam Black, who shared with me an undergraduate paper he wrote entitled "Manufacturing Authenticity Levi's® Vintage Clothing: A Jeaneology."

106 *The absolute champ of working-class denim authenticity*: Rob Walker, "Jeans Engineering," *New York Times Magazine*, August 28, 2005, www.nytimes.com/2005/08/28/magazine/28CONSUMED.html.

107 *the driving force behind the* Encyclopédie: Philipp Blom, *Encyclopédie: The Triumph of Reason in an Unreasonable Age* (New York: Fourth Estate, 2004).

108 *The German version became a sensation:* Trilling, *Sincerity and Authenticity*, 27.

109 *"a compound of elevation and abjectness":* Denis Diderot, *Rameau's Nephew and Other Works*, trans. Jacques Barzun and Ralph H. Bowen (New York: The Library of Liberal Arts, 1964), 8–9.

109 *In his introduction to his own translation:* Ibid., 7.

109 *As Diderot lectures the nephew:* Ibid., 83.

111 *If we have trouble following Hegel:* Trilling, *Sincerity and Authenticity*, 36.

112 *"an architect of consumer dissatisfaction":* Joshua Glenn, "Fake Authenticity: An Introduction," *Hermenaut*, December 22, 2000, www.hermenaut.com/a5.shtml.

113 *Bullshit makes no such acknowledgment:* Harry G. Frankfurt, *On Bullshit* (Princeton: Princeton University Press, 2005), 65.

116 *"I think people are craving the earth":* David Gelles, "Down and Dirty," *The New York Times*, February 8, 2007, www.nytimes.com/2007/02/08/ garden/08dirt.html?_r=2&pagewan.

116 *Veblen's account of the goals and motivations:* The discussion of Veblen in this chapter (as well as the term "conspicuous authenticity") is heavily indebted to Joseph Heath. See Joseph Heath, "Thorstein Veblen and American Social Criticism," *Oxford Handbook of American Philosophy*, ed. Cheryl Misak, (Oxford: Oxford University Press, 2008).

117 *Veblen remarks that the transition:* Thorstein Veblen, *The Theory of the Leisure Class* (New York: The Modern Library, 2001), 220.

121 *The rebel consumer goes to great lengths:* For a full discussion of rebel consumerism, see Joseph Heath and Andrew Potter, *The Rebel Sell: Why the Culture Can't Be Jammed* (Toronto: HarperCollins, 2004).

128 *Simply feeding that many cattle:* Alex Avery, *The Truth About Organic Foods* (Chesterfield, MO: Henderson Communications, 2006). See also Louise Gray, "Organic Food Has No Added Nutritional Benefit, Says Food Standards Agency," *The Daily Telegraph*, July 29, 2009, www.telegraph.co.uk/earth/earthnews/5932100/Organic-food-has-no-added-nutritional-benefit-says-Food-Standards-Agency.html.

128 *The first group reported much higher levels of appreciation:* Hilke Plassmann, John O'Doherty, Baba Shiv, and Antonio Rangel, "Marketing Actions Can Modulate Neural Representations of Experienced Pleasantness," Proceedings of the National Academy of Sciences, January 14, 2008, www.pnas.org/content/105/3/1050.full.

128 *"the magic cure-all, synonymous with eating well":* Mark Bittman, "Eating Food That's Better for You, Organic or Not," *The New York Times*, March 22, 2009, www.nytimes.com/2009/03/22/weekinreview/22bittman.html.

130 *The environmental benefits of local farming are actually highly overstated:* Drake Bennett, "The Localvore's Dilemma," *The Boston Globe,* July 22, 2007, www.boston.com/news/globe/ideas/articles/2007/07/22/ the_localvores_dilemma.

131 *This is all part of one of the greatest fads:* Elizabeth Kolbert, "Green Like Me," *The New Yorker,* August 31, 2009, 70–74. For the story of The Compact, see Zachary Slobig and Dan Hennessy, "The Compact," *Good Magazine,* October 3, 2007, http://www.good.is/post/the-compact/.

CHAPTER 5

138 *"Does authenticity still matter?":* Richard Siklos, "I Cannot Tell a Lie (from an Amplification)," *The New York Times,* February 5, 2006, www.nytimes.com/2006/02/05/business/yourmoney/05frenzy.html.

141 *Any hopes that it might be otherwise:* J.M. Christensen-Hughes and D.L. McCabe, "Understanding Academic Misconduct," *Canadian Journal of Higher Education* 36:1 (2006): 49–63.

141 *Lest anyone think that Canadian students:* See the studies tracked by Professor D.L. McCabe of the Center for Academic Integrity at www.academicintegrity.org.

141 *But what is plagiarism exactly?:* Richard A. Posner, *The Little Book of Plagiarism* (New York: Pantheon, 2007).

143 *The more accomplished the product, the more accomplished the maker:* André Gombay, "'The more perfect the maker, the more perfect the product': Descartes and fabrication," *Philosophy,* 71:277 (July 1996): 351.

153 *As law professor and copyright reform activist:* Lawrence Lessig, *Free Culture: How Big Media Uses Technology and the Law to Lock Down Culture and Control Creativity* (New York: Penguin, 2004).

155 *Say good-bye to today's experts:* Andrew Keen, *The Cult of the Amateur: How Today's Internet Is Killing Our Culture and Assaulting Our Economy* (London: Nicholas Brealey Publishing, 2008), 9.

156 *But Sunstein is worried that group polarization:* Cass Sunstein, "How the Rise of the Daily Me Threatens Democracy," *Financial Times,* January 11, 2008, A9.

158 *In early 2008, the Project for Excellence in Journalism:* See the report at www.journalism.org.

159 *"the triumph of innocence over experience":* Keen, *The Cult of the Amateur,* 36.

160 *Thanks to the Panopticon, he wrote:* Jeremy Bentham, "Panopticon," The Panopticon Writings, ed. Miran Bozovic (London: Verso, 1995).

163 *"gossip is inherently democratic":* Ronald de Sousa, "In Praise of Gossip: Indiscretion as a Saintly Virtue," *Ethics for Everyday,* ed. D. Benatar (New York: McGraw-Hill, 2002), 117–125.

CHAPTER 6

170 *Forty-seven years later, a Harvard business professor:* John Quelch, "The
    Marketing of a President," *Harvard Business School Working Knowledge,*
    November 12, 2008, http://hbswk.hbs.edu/item/6081.html.

171 *"The qualities which now commonly make a man":* Joe Klein, *Politics Lost:
    How American Democracy Was Trivialized By People Who Think You're
    Stupid* (New York: Doubleday, 2006), 22.

171 *As* Time *magazine columnist Joe Klein puts it:* Ibid., 23.

172 *As Rick Perlstein puts it in* Nixonland: Rick Perlstein, *Nixonland: The Rise of
    a President and the Fracturing of America* (New York: Scribner, 2008), 58.

174 *"The Spectacle is not a collection of images":* Guy Debord, *The Society of
    the Spectacle,* trans. Donald Nicholson-Smith (Cambridge: Zone Books,
    1995), Thesis 4.

175 *"On the 26th of July, which out in Missouri we call Turnip Day":* Klein,
    Politics Lost, 22.

176 *"In the process," writes Klein:* Ibid., 23.

176 *In the concluding pages of his* Politics Lost: Ibid., 240.

187 *Always willing to trade economic hope:* Thomas Frank, *What's the Matter
    with Kansas: How Conservatives Won the Heart of America* (New York:
    Metropolitan, 2004).

189 *But as Dahlia Lithwick wrote:* Dahlia Lithwick, "Lost in Translation,"
    Slate, July 8, 2009, www.slate.com/id/2222523/

193 *Almost every losing party or candidate eventually gets desperate:* Dana
    Milbank and Jim VandeHei, "From Bush, Unprecedented Negativity,"
    The Washington Post, May 31, 2004, A1.

193 *According to a study by Brown University professor:* Darrell West, *Air
    Wars: Television Advertising in Election Campaigns, 1952–2004, 4th Edition*
    (Congressional Quarterly Press, 2005).

194 "People always say they hate it when newspapers print photographs":
    Warren Kinsella, *The War Room: Political Strategies for Business, NGOs,
    and Anyone Who Wants to Win* (Toronto: Dundurn, 2007), 159.

194 *A lot of political advertising gets labeled "negative" when it isn't:* Ibid., 167.

196 *a successful brand can actually create a niche:* Al Ries and Laura Ries, *The
    22 Immutable Laws of Branding: How to Build a Product or Service into a
    World-Class Brand* (New York: HarperCollins, 1998).

197 *negative campaigning should increase voter turnout:* Peter Loewen, "Affinity,
    Antipathy, and Political Participation: How Our Concern for Others
    Makes Us Vote," forthcoming in the *Canadian Journal of Political Science.*

197 *The fact that German society was already so heavily politicized:* Richard J.
    Evans, *The Coming of the Third Reich* (New York: Penguin, 2003), 118.

CHAPTER 7

201 *As a Pacific Island dancer replied:* Denis Dutton, "Authenticity in Art,"
The Oxford Handbook of Aesthetics, ed. Jerrold Levinson (New York:
Oxford University Press, 2003), www.denisdutton.com/authenticity.htm.

202 *Instead, it becomes a destination almost exclusively for tourists:* Ibid.

204 *which we can call a "worldview" or "ethos":* Tyler Cowen, *Creative
Destruction: How Globalization Is Changing the World's Cultures*
(Princeton: Princeton University Press, 2002), 51.

205 *the anxiety of influence:* A tragic version of this is novelist David Foster
Wallace, who suffered from what is likely the most severe case of influ-
ence-anxiety of all time. He was so aware of his influences (writers
such as Thomas Pynchon and Don DeLillo) that he larded his books
and essays with footnotes, parenthetical remarks, quotation marks, and
so on, lest the reader conclude that any of what Wallace had written
was original. He killed himself in 2008.

206 *Noting the proliferation of religious sects:* Roy Porter, *Enlightenment:
Britain and the Creation of the Modern World* (London: Penguin, 2000),
108.

207 *As social theorist Grant McCracken argues:* Grant McCracken, *Plenitude:
Culture by Commotion* (Toronto: Periph: Fluide, 1997).

208 *As German philosopher Christoph Wieland wrote:* Kwame Anthony
Appiah, *Cosmopolitanism: Ethics in a World of Strangers* (New York:
Norton, 2006), xv.

208 *This view was echoed a century or so later:* John Stuart Mill, *On Liberty*
(Oxford: Clarendon Press, 1980), 125.

211 *As Kwame Anthony Appiah puts it:* Appiah, *Cosmopolitanism*, 204, my
emphasis.

211 *The principles/values distinction is one that we don't usually make:* Joseph
Heath, *The Myth of Shared Values in Canada* (Ottawa: Canadian Centre
for Management Development, 2003).

212 *the work of American sociologist Robert Putnam:* Robert Putnam, "E
Pluribus Unum: Diversity and Community in the Twenty-first Century
– The 2006 Johan Skytte Prize Lecture," *Scandinavian Political Studies*
30:2 (2007): 137–174.

217 *It is easy to see how this same pattern of explanation:* This discussion is
drawn from Fred Hirsch, *Social Limits to Growth* (Cambridge: Harvard
University Press, 1976), chapter 5.

220 *a necessary evil of urban life:* For an exceptionally grim description of
this, see Luc Sante, *Low Life: Lures and Snares of Old New York* (New
York: Farrar, Straus, & Giroux, 2003).

222 *"a multitude of uniform, unidentifiable houses":* Lewis Mumford, *The City in History: Its Origins, Its Transformations, and Its Prospects* (New York: Harcourt, Brace & World, 1961), 486.

224 *"When Americans, depressed by the scary places":* James Howard Kunstler, *The Geography of Nowhere: The Rise and Decline of America's Man-Made Landscape* (New York: Touchstone, 1994), 185.

225 *The elites are not interested in hearing:* Robert Bruegmann, *Sprawl: A Compact History* (Chicago: University of Chicago Press, 2005), 161.

227 *the trade in cultural goods and services, tends to have the following effects:* The rest of this section draws heavily on Cowen's *Creative Destruction,* chapter 5.

CHAPTER 8

235 *"the psychological Sahara that starts right in your bedroom":* Joseph Brodsky, "In Praise of Boredom," *Harper's Magazine,* March 1995.

236 *He argued that the relentless extension:* Francis Fukuyama, "The End of History?" *The National Interest,* Summer 1989, http://www.wesjones.com/eoh.htm. See also Francis Fukuyama, *The End of History and the Last Man* (New York: Free Press, 1992).

239 *"Some like pizza, some like steaks":* George Grant, *Technology and Empire* (Toronto: House of Anansi, 1969), 26.

240 *"The end of history will be a very sad time":* Francis Fukuyama, "The End of History?"

244 *As Anne Applebaum writes:* Anne Applebaum, *Gulag: A History* (New York: Anchor, 2003).

251 *Mussolini in particular liked to brag:* R.J.B. Bosworth, *Mussolini* (London: Arnold, 2002).

251 *In the end, what defines fascism in all guises:* Robert O. Paxton, *The Anatomy of Fascism* (New York: Vintage, 2004), 219.

252 *The nostalgia for the past unity:* Paul Berman, *Terror and Liberalism* (New York: Norton, 2003), 47–49.

254 *And so, as Lawrence Wright puts it:* Lawrence Wright, *The Looming Tower* (New York: Knopf, 2006), 24.

254 *Indeed, the man in the cave:* Wright, ibid., 235.

259 *Yet on her trip, Wolf discovered:* Naomi Wolf, "Behind the Veil Lives a Thriving Muslim Sexuality," *The Sydney Morning Herald,* August 30, 2008, www.smh.com.au/cgi-bin/common/popupPrintArticle.pl?path=/articles/2008/08/29/1219516734637.html.

261 *Amis struggled more than most to find his footing:* Martin Amis, *The Second Plane: September 11: Terror and Boredom* (New York: Knopf, 2008).

262 *"The age of terror, I suspect, will also be remembered":* Ibid., 76.

CONCLUSION

265 *In his book* Unpopular Essays, *philosopher Bertrand Russell:* Bertrand Russell, *Unpopular Essays* (New York: Simon and Schuster, 1964), 146.

266 *There are no real leaders in a liberal state:* This quotation, along with much of the discussion here, is drawn from Francis Fukuyama, *The End of History and the Last Man* (New York: Avon Books, 1992), 300–321.

266 *Nietzsche's contempt for the morality of liberal democracy:* Ibid., 304.

269 *"Reflecting on what we leave to our grandchildren":* David Suzuki with Faisal Moola, "A Grumpy Old Man Ponders the Past," *Toronto Sun*, February 18, 2009.

INDEX